WITHDRAWN

SCOTTISH BORDERS LIBRARY SERVICES	
009377033	
Bertrams	28/04/2014
600	£16.99

the book

CONTENTS

FEATURES

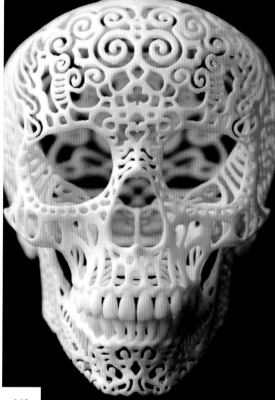

062

142

150

130

Hello and welcome to The Gadget Show Book

Wow, a Gadget Show book! Not since *Look-In* and *Top of the Pops* (my two favourite annuals from the 1970s and '80s) have I been so excited about a new book! So a big welcome to our book. Much of the awe-inspiring tech and innovations that *The Gadget Show* has explored during the last seventeen series are contained within these pages. The birth of 3G, the arrival of WiFi, the launch of the smart phone and the million-seller tablet, well none of these things existed in any concrete way when we filmed the very first episode of the show in a small car park in Digbeth, Birmingham.

We compare the facts and dispel many of the myths behind some of the most iconic tech ever invented, and offer more top tens and historic tech face-offs than you can shake an accelerometer equipped gaming controller at.

While on the program we have tried to steer clear of making wild tech predictions for fear of getting egg on our faces (I am still convinced my Sinclair C5 is the future of transport!), but in the book we look at the tech that was accurately predicted in movies.

This book shouldn't just inform, it should also entertain, inspire, and excite – qualities that we try to sprinkle over every episode of the TV show. Roll on the next decade of the show, because if it is anything like the last, we're in for a real tech-laden treat. So sit back, relax and grab yourself an extra shot of Mocha-Choco-Cappuccino-Latte and get ready to satisfy your inner geek.

Jason

SCOTTISH BORDERS COUNCIL

LI RA Y &

INFORMATION SERVICES

I guess you could call me the
Crazy Gadget Traveller!!! I spend more
nights in hotels around the globe than I do at home
. . . but LOVE IT! While testing tech, thrill-seeking,
following the golf and blogging, I capture my travels
through the lens on my CSC. I know what you're
thinking . . . and
YES . . . you can fit in my case. xx ☺

Polly

People constantly tell me that
working on *The Gadget Show* must be one of the
best jobs in the world. They're absolutely right. It's
a great privilege to test and enthuse about gadgets for a
living. In this book Rachel, Polly, Jason and I aim to share
our delight in the ever-changing, interminably fascinating
gadget world. I sincerely hope this book helps you have
even more fun with yours!

Jon

Hello from the new girl! With the first series of *The Gadget Show* aired and the second series in the can, I can say it's been an absolute blast. Working with Jason is brilliant and getting to experience at first-hand the best tech the world has to offer is this geeky girl's dream come true. Hope you enjoy this book as much as I enjoy making the show.

Rachel

POLLY'S TECCHIE THRILLS DIARY

All action-girl Pollyanna spent the last series of The Gadget Show travelling the world to test the best adventurous tech there is. She drove, flew, crashed and broke (yikes!) all manner of mind-boggling stuff. Here is Polly's diary of her very first Polly's Tecchie Thrills trip to North America.

DAY 1

For me series 17 of *The Gadget Show* begins today. My exciting new part of the show is called *Polly's Tecchie Thrills* and it will see me travelling the globe to test, ride, drive, fly and master (while sometimes picking up the odd bruise) the craziest new inventions the tech world has to offer. My first trip (across America and Canada) will include five shoot days and three travel days; I will take seven different flights and will cover a total of 6500 miles by air and road. What I will be testing though, remains — for the time being — unknown. I am at the mercy of my production team; it seems

Cameraman Chris, me, producer Alex and soundman Ben assemble in San Francisco for our first *Polly's Techhie Thrills* North America trip. Yippee!

Jason Woods and his sadly never really got it working Kymera Body Board. Pic below – At the controls of the insane Seabreacher X. Loved it!

that the 'who' and 'what' are for them to know and for me to find out! So, with a week's worth of clothes, my phone, a compact system camera (CSC) and a plane ticket for San Fran, I head off for the west coast of the USA.

After two hours I touch down in SF and take a one and half hour cab ride to Napa, finally arriving at our 'somewhat basic' roadside hotel. I soon spot the crew: director Alex, cameraman Chris and soundman Ben6. They flew in on an earlier flight from London and are still busy un-boxing and setting up the mountain of kit for our first shoot tomorrow. After a quick natter (including finding out it took them just 30 minutes to receive a toll booth fine) I hit the hay as the boys head off for a late (10pm) dinner! Chris texts to say his crispy orange chicken was a mistake!! It should be a great car jpurney tomorrow!

DAY 2

My day starts in a Denny's restaurant next door. I walk in to find the boys demolishing three gargantuan platters of eggs, bacon and waffles. So after joining them, we load up our ride, the truly beastly GMC Yukon 4x4. Now while this car might have nine seats, five of them are full with kit bags – no such thing as travelling light when filming! The drive is 50 minutes to Lake Berryessa, the biggest lake in Napa County; and if the tree-lined approach wasn't beautiful enough, the lake itself is truly breathtaking. There we meet up with Jason Woods, the inventor of the Kymera Body Board: an electric lie-on watercraft powered by a 2000W brushless motor that is capable of a fifteen mile per hour top speed.

Things start well and I quickly get to grips with the Kymera, cutting through the windswept, crystal blue waters. However, as I decide to open it up, disaster strikes. The engine dies and I am forced to paddle back to shore. Jason diagnoses a problem with the speed control module and thankfully manages to fix it. So soon I head back out on the water to try again. But, less than five minutes later, I'm swimming doggy paddle back to shore. This time the Kymera has well and truly died and is beyond repair.

Testing over after just fifteen minutes! Funny thing is, we had hired a boat to shoot high speed tracking sequences later in the day, so Alex is having a fun time in the shop asking for our money back! Good luck with

Our 5.3 litre V8 GMC Yukon 4x4. What a monster! Pity it was a parking ticket magnet

The Seabreacher X is an insane ride. I can't wait to take over the controls.

that (lol)! Having said our goodbyes to Jason we drive around the lake to find a secluded spot (a quieter spot for our soundman Ben who is serioulsy anal about silence (lmao)!!) to record my post-match interview. During this Alex and Chris spot an American Eagle which Ben misses. Needless to say, they won't let him forget it for the rest of the day! After lunch we hit the road, heading north for Redding. Four hours — and eight Almond Snickers They are soooo good) — later we arrive. Fact of the day? Lake Berryessa is home to mountain lions.

DAY 3

Another day, another lake. This time Shasta Lake, California's largest reservoir. It's stunning, but unlike yesterday, it is hot — 35-degrees centigrade hot. And today's technology is equally scorching. I am here to try out two machines created, designed and built by Kiwi Rob Innes and his team at Innespace Productions.

First is the Seabreacher X: a $90,000, 260 horsepower, 50mph shark-shaped boat. Capable of gliding through, diving under and rising out of the water: this thing is insane. So, with the crew strapped into a support filming-boat, and me (plus two GoPro cameras) positioned in the back seat of the Seabreacher, Rob shows me the ropes.

Now trust me, nothing can prepare you for a ride like this. It is fast and frantic — turning on a sixpence and slicing through the water as it goes! And the jumps? Well they're insane — six foot

Finally got the hang of it! And my water phobia is a thing of the past, for today at least.

high and almost vertical. The lump on the top of my head proves just how much this thing throws you around! Rob then lets me to take the controls, so after a mid-lake seat swop I begin my attempts at dives and jumps. It is testament to the precision engineering that even a novice like me can drive it without any training. I pull a mini-jump as well as mastering the submersible dive. But as if that isn't enough fun for one day, Rob has another surprise in store.

This seriously creeped me out!

Enter the Jetovator, best described as a high-powered water bike. Water expelled from an attached jet ski is forced up a hose and out of two high-pressure jets on the bike's handles. Moving them around gives you the ability to change height and direction. Rob's team make the Jetovator look easy, but I am terrified. I have a serious water phobia, and any gadget that involves me being submerged repeatedly is always going to be right up there on the 'get me out of here' (once you've got the shots!)scale. So, to be honest I wasn't over the moon about this. However, with encouragement from everyone there, I don my helmet and slide gingerly into the water. Rob slowly builds up my confidence and to my amazement I soon emerge from the water atop the Jetovator. It may have been a two-foot flight, but I loved it, and I want to go again.

What follows is half-an-hour of adrenaline-pumping Jetovator-taming action. It's incredible — an experience like no other. As well as taming the gadget, I have — for one day at least — overcome my dislike of water. As we return to dry land, I can see the looks of ill-concealed jealousy on the crew's faces and it is at times like this that I realize that I am ticking off my dream bucket list. Having loaded up the Yukon and rolled out of the car park — complete with an unwelcome parking ticket — we head north for Grant's Pass, a small town in Oregon. Before long we make a well-deserved pit stop for snacks and drinks. Dehydrated and hungry after an exhausting day's filming in the sun, we pull over at the first available service station.

Now, this isn't your average BP Garage with a Costa attached. This is genuine middle-America, back-of-beyond territory. Inside the wooden hut we find what we need in the way of sustenance (ice creams, drinks, sweets and apple for good measure), but it's when the owner invites us to 'explore the back of the shop' that we are in for a completely surreal experience.

Fun, but not sure it will ever really catch on . . .

First we come across a four-foot python (yikes!), and in the bathroom, a female mannequin, fully dressed, lying casually in the bath tub! Weird doesn't begin to describe it, but memories are made of this.

DAY 4

After breakfast I step outside to find the crew in hysterics. Having caught too many rays yesterday, Ben has gone mad with the sun cream this morning and is doing a good impression of Caspar The Friendly Ghost (although nowhere near as cute)! We are here in the quaint town of Grants Pass to meet local inventor Charles Greenwood and his creation: The Human Car. This is basically a rowing machine with a carbon fibre car body wrapped around it. But before we can start filming we are met by Charles's entourage — his son, business partner, cameraman, soundman, truck driver and head engineer. Suddenly a quiet street has just got very busy indeed.

As our filming begins, we are surrounded by even more of the Grant Pass population, all leaning in for pictures and shouting questions at us without a care in the world. Charles and I decide to make a move, rowing ourselves onto the high street, weaving through four lanes of traffic and waving at bemused onlookers as we got.

The Human Car — though not that practical — is great fun, and I love the open-air bicep workout. I even up the ante

Chris doing what he does best.

The 135km long Okanagan Lake. It's supposed to be the home of a serpent-like monster called Ogopogo, but we never saw it, sadly!

and persuade two buff guys to come out of the local gym to help Charles and me increase the rowing pace, something that will also help charge the Human Car's battery! We get up to 30mph on the open pine tree-studded country roads and the wind-rush feels amazing.

As if we hadn't turned enough heads already, we are soon being tracked by the local news team as a 'fresh-out-of-college' reporter hops out of one of those quintessential American news trucks to interview Charles's brother. Our work is done and we leave the 'local heroes' to it. We jump into our now-beloved Yukon, crank up the aircon and hit Interstate 5 heading due north to Portland (Oregon). Chris is driving (playing the 'how long can I drive on the rumble strips for' game); Alex is on DJ duty and we are all belting out our best rendition of 'Sweet Home Alabama'. Four hours later – having passed through Lebanon, Lookingglass and Sweet Home (can you believe it?) – we roll into the nighttime lights of Portland. As the boys unload the kit into the hotel, the porter reveals that 'The Goonies' was filmed nearby. Pity we won't have a chance to visit the set. The boys settle for a round of Blue Moon and I hit the 24-hour gym!!!

DAY 5

As my cab pulls up at Portland International Airport I spot my faithful trio at the front of the check-in queue. Turns out they've been here for two hours already to return our Yukon, get their carnet stamped (an official document required when transporting commercial equipment through

Chris (again) chills after all the hassle at Vancouver airport

airports), and to drop off our mountain of black Peli Cases. Today we are crossing the border into Canada. Our destination is Kelowna in the Okanagan Valley, but first we must transit through Vancouver.

Director and our 'dad for the week' Alex, pulls out the pocket-money and buys lunch at a Hawaiian themed restaurant complete with grass skirted waitresses and beach hut seating booths

My first sight of the Maverick Flying car. I can't wait to try it!

(only in the US!). Going through security with a film crew is always amusing. I breeze through and look back as the boys are being given the official once over. Alex and Chris's cameras are being swabbed, Ben's sound kit is being pulled apart much to his distress, and their visas are being thoroughly investigated as ever. The only thing missing is a strip search.

Two hours later, having landed in Vancouver, it's the same story. Customs take one look at our piled up kit and ask to see us 'in the back'. We oblige, so we haul our bags on to the counter and the examination begins again. Having rifled through all our bags, paperwork and IDs we are liberated and wished 'A Great Day'. We rush through to departures, clear security (again) and board our 50-seater Air Canada Dash 8-300 plane to Kelowna. At 11.00pm we land, load up our bright red hire van and head thirty minutes north right into the back of beyond. Oh and it's raining, pouring actually. This doesn't bode at all well for tomorrow's filming.

Time for my chat with Rachel and Jason back in The Gadget Show studio.

DAY 6

My alarm penetrates my sleep-befuddled brain at an unearthly 5.30am, and by 6.00am we are on the road again (yes I can get ready that quickly!); I munch on a takeaway breakfast bap supplied by the hotel. Fifteen minutes later we pull into Vernon Airport, where Ray Siebring is waiting for us. Ray is the owner of a Maverick Flying Car. Yes, that's right — a car that flies! Suddenly my early rise

Boys will always be boys, I guess!

is forgotten and I am now champing at the bit to see this beauty in action.

However, as gusts of wind whip across the airfield, Ray's face is a mask of furrowed brows and evident concern. While we begin to unfurl and hoist the parachute that allows the Maverick to fly, Ray tells me that flying will not be possible if the wind speed exceeds seven knots (approximately eight mph). Early mornings usually mean calmer conditions, but today the wind speed is right on the upper limit. Half an hour of consultation follows until finally ... yes, it is go time! Ray straps me into the back of the Maverick, Alex and Chris start the GoPro cameras recording and then jump aboard a Cessna 172 tracking plane. Now I am strapped in I can't help feeling a stomach-swirling mix of emotions ranging from the deeply apprehensive all the way through to absolute ecstasy. Ray fires up the rear propeller, air fills the parachute above and we roar down the runway. After just 80 metres we hit the magic take-off speed of 40mph and the wheels separate from the tarmac. We are airborne! For a RARE moment I am speechless. The overwhelming beauty of the experience is incomparable; I am alive with giddy excitement and a sense of reverence along with awe and reverence. To my right I spot Alex and Chris flying and filming alongside; to my left is the ever-shrinking airport and ahead the glistening blue waters of the Okanagan Lake.

The next hour in the skies creates enough memories to last a lifetime. This is the kind of thrilling experience I will talk about for the rest of my life. I am not ashamed to admit that I loved the Maverick Flying Car and the experience so much that it brought a tear to my eye. Back on terra firma we express our gratitude to Ray and bundle into our altogether more boring hire car.

We have to make another flight, so step on it as we head for Vancouver Airport. To pass the five-hour journey, I take the wheel and beat Chris's rumble strip straddling record, Ben pumps the tunes and we chow down on bumper portions of well-deserved McDonalds. [Hey, here's some trivia for you: Canada is the only country in the world that has a modified McDonalds logo. The golden 'M' has a tiny red Canadian Maple Leaf in it!].

Business class grub. Grrrr! All I got was a cellophane wrapped sandwich and a tub of water.

You know what, the snow-capped mountains and tree-lined gorges makes Highway 1 a bloomin' beautiful place to be on a sunny day like today. One problem though — a severe lack of service stations. So, the phrase, 'when you gotta go, you gotta go' is the order of the day. Bless those boys!

DAY 7

Our quartet reunites in the queue to board United 246 to Chicago's O'Hare Airport. It quickly transpires that some of us are going to enjoy the next five hours more than others. Alex, Chris and Ben (who arrived earlier to clear customs) have been unexpectedly upgraded to business class, while I have an economy ticket. Alex offers to swap seats, but I am happy to let the boys stick together and enjoy their prawn cocktails, chicken burritos and cheesecake (see opposite). On arrival in Chicago, I make a beeline for lunch, but the boys in tow aren't hungry for some reason! Once we're all on a sustenance even keel, we take yet another flight, this time bound for Portland (Maine) two hours away. We are back in the good ol' US of A — this time on the northeastern coast — to hook up with two very clever twin brothers. Mike and Geoff Howe are the founders of, and innovators behind, Howe and Howe Technologies based in Waterboro, where their core business is designing and building rugged vehicles and robots for the military. I am met by the boys and driven at break-neck speed to their workshops in a

Now this is what I call a serious bit of tech!

This was one challenge I managed to rise to!

modified 800bhp Chevy Camero (an absolute beast of a car!). I am here to see three of their latest creations: the SwatBot robotic ballistic shield; the Thermite remote control firefighting robot and the imposing RipTide. The latter is a bespoke tank-track vehicle that was originally designed for Dwayne 'The Rock' Johnson to use in the new GI Joe 2 blockbuster movie.

It is as thrilling as a vehicle gets, flinging me about from side to side as we dive into water, tear up dirt tracks and plough through mounds of debris. We push it so hard that the engine blows up at the end! I call that a thorough testing or just the Polly Touch (lol)! Shoot over we are given a tour of the inner sanctum of the boys' world and the pick of their own TV show's official clothing line; are shown other amazing Howe and Howe creations; and dine on mega Maine burgers. That's just how we roll in the world of TV.

DAY 8

Today we are homeward bound. The first ever *Polly Tecchie Thrills* filming trip is complete. On the way to the airport Chris buys a cut-price iPad from Best Buy and Alex gives the 'Rumble Strips' game his best shot. At Portland Airport's check-in desk we say our goodbyes as I off to Phoenix, while the boys are making tracks to London via New York.

When I land I get a call to tell me they are having an eventful journey home. It turns out that their flight to New York was delayed by four hours (so plenty of time to spend meditating on life on Portland Airport's rocking chairs!) and consequently they have missed their connecting flight to London. As I write this, I understand they are holed up in Ramada three miles from Newark Airport eating pizza, sharing rooms and facing the prospect of getting up for the early flight to London. That's a wrap for now folks.

Chris really does know how to chillax anywhere, any time!

AUDIOVISUAL
ULTIMATE GADGET FACE-OFFS

Technology that uses sight or sound (or a combination of both) to relay information to the user is something that we take for granted nowadays. If Sony hadn't come up with the original 'wired for sound' cassette Walkman, we might not be so routinely plugged into our iPods loaded with our favourite playlists. If VHS video hadn't revolutionised the way we watch film, then DVDs, Blu-ray and on-demand streaming might never have developed in the way they have.

The technology may have become more sophisticated, with staggering improvements in sound and image quality, with data storage that's growing exponentially and can be accessed at blindingly fast speeds, but at the heart of all the clever tech, we still value the nature and quality of the content above all else.

In this section we pit the originators of each type of media 'delivery system' against their modern tech succesors in the ultimate audio/visual face-off.

IPOD

Success: 5
On 9 April 9 2007 Apple sold its 100 millionth iPod, making it the fastest selling music player in history.

Lifetime: 3
2001 to now.

Innovations: 4
The iPod was the first music player that allowed users to store multiple albums.

Fun Facts
It's a life saver! An iPod was vital in keeping a patient conscious during a case of brain surgery. It also saved a soldier's life by blocking a bullet when he was shot in the chest.

VHS AND DVD

Success: 5
VHS was launched in 1976 and, after the format war with Betamax, became standard home format by the 1990s.

Lifetime: 5
37 years (VHS launched in 1976, DVD still going).

Innovations: 4
For the first time it meant people could watch recorded television programmes and commercial films when they wanted, liberating them from the schedulers and cinema.

Fun Facts
DVD is not an actual acronym. When asked, people in the industry will say it stands for 'digital versatile disk'; however the launch press releases and subsequent marketing materials make no actual mention of what the letters referred to.

WALKMAN

Success: 4
The Walkman has sold over 200 million units in all guises since launch.

Lifetime: 5
1979 to 2010 (cassette playing version).

Innovations: 5
The Walkman was the first mass-produced small portable music player that allowed you to carry your own music with you, although a portable personal player called the Stereobelt had been patented in Europe.

Fun Facts
In 1986 the word 'Walkman' appeared in the Oxford English Dictionary for the first time.

LOVEFiLM.COM®

ONLINE STREAMING

Success: 4
In 2012 streaming joined DVD rental at Lovefilm and it is only a matter of time until it entirley supplants physical distribution of music and film.

Lifetime: 3
2006 to now.

Innovations: 4
For the first time the service gives access to a huge library of media without the need to visit shops or libraries. It is rapidly changing the purchasing habits of consumers and is ushering in a host of new business models.

Fun Facts
2012 data shows that in North America, during peak hours (9pm to 12am) audio and video streaming now takes up around 65% of all Internet traffic.

SUPER 8

Success: 4

Created a visual style and look that is still used today in music videos and art-house films with pretentions.

Lifetime: 5

Still in use today; there are special film festivals dedicated to Super 8.

Innovations: 4

The first consumer movie camera to use film cartridges. Owners avoided the fiddly need to load film. It was effectively an early plug and play gadget.

Fun Facts

The characteristic look of the resulting film owes a great deal to the camera's eighteen-per-second frame rate.

BOX BROWNIE

Success: 5

Launched in February 1900 by Eastman Kodak, this relatively inexpensive camera sold 150,000 units in the first year alone. The 127 model launched in 1952 would sell quite literally in the millions.

Lifetime: 5

70 years.

Innovations: 4

The Brownie shifted the perception of photography in popular culture. Until its arrival photography was a cumbersome technical affair; thereafter it became one of the most popular and enduring activities in most families in the western world. Eastman Kodak's sales slogan: 'You Push The Button, We Do The Rest' was transformative.

Fun Facts

The Brownie name was derived from an immensely popular series of verse cartoon books and comic strips featuring mischievous kindhearted fairies called The Brownies by Canadian artist Palmer Cox.

SPOTIFY

Success: 5

Took just two years from launch to reach ten million users. Now has a music catalogue of more than twenty million tracks.

Lifetime: 3

From 2006 to now.

Innovations: 4

Now you can stream music via computer, tablet and smart phone for free from a huge and growing online library. A truly revolutionary way to consume music.

Fun Facts

The Beatles are not available because of a digital distribution agreement exclusive to iTunes.

SD CARDS

Success: 4

Rapidly established as the main choice card for data storage in digital cameras.

Lifetime: 4

1999 to now.

Innovations: 3

The popularity of this storage device for digital cameras was driven by its size and convenience; and the way it allowed for relatively easy transfer of recorded images from camera to computer.

Fun Facts

Interestingly the SD logo was originally designed for another purpose altogether. Toshiba planned to enter the DVD format battle with a Super Density storage disk. Once the project was abandoned the Secure Digital card inherited the acronym.

POLAROID

Success: 4

By 1962 the number of Polaroid camera owners worldwide exceeded four million.

Lifetime: 4

1948 to now – despite the announcement by Polaroid in 2008 that production of the instant film would cease. A decision that was reversed a year later.

Innovations: 4

The first camera that actually developed photos within camera –and in an incredible sixty seconds or so.

Fun Facts

In the years since the Polaroid camera was launched, at any given moment more than one hundred Polaroid pictures are developing in users' hands across the world.

ITUNES

Success: 5

Apple launched iTunes in 2003, and within five years the service had overtaken sales of music CDs. Ten years later in February 2013, Apple announced that it had sold over 25 billion songs via the iTunes Store. This means it is averaging 15,000 downloads-per-minute from its 26-million plus song catalogue that is available in 119 countries.

Lifetime: 3

From 2003 until now.

Innovations: 5

The first (legal) downloadable record store on your computer.

Fun Facts

Apple gave American Ales Ostrovsky an iMac, 10 fifth generation iPods and a $10,000 gift card when he made the billionth iTunes download. His choice of download? Coldplay's 'Speed of Sound'.

UTTERLY STUPID
GADGETS

10 OF THE ~~BEST~~ WORST

Over the years we have had a few tech count downs with a difference, and this list is always top of our 'what on earth were they thinking?' discussion when we are gathered around the Gadget Show ideas table. Even though much of the actual technology and concepts behind the products were perfectly well intentioned, the final execution often left much to be desired. So here is our Top 10 all time favourite moments of gadget lunacy.

NOT WANTED

Motorised Twirling Ice Cream Cone
⑤ £4.99 (best price)

The Motorized Ice Cream Cone. It Allegedly Reduces 'Drippy Cone Syndrome'.

Forget wafer, waffle, chocolate or sugar dipped – doesn't dayglow plastic fire up your appetite for a nice artisan dairy ice cream? And, as the advertising blurb that comes with it boasts, 'No more licking around the edges of a drippy cone.' Yeah, right! Frankly we can't see the point when half the enjoyment of an ice cream cone is eating it at the end. Just the thought of this pointless culinary conception is enough to set your teeth on edge!

NOT WANTED

The annoy-a-tron. 'Beep, Beep, Beep.' A Truly Annoying Little Tool.

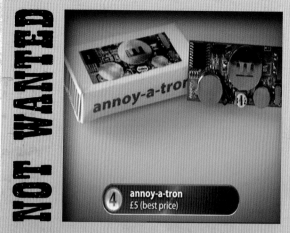

annoy-a-tron
£5 (best price)

Anything with the word 'tron' in its title should be cool right? But this is definitely the exception to the rule. Instead of futuristic CGI-laden entertainment, this tiny battery driven circuit is responsible for merely generating a short and very annoying beep every few minutes at either a user selectable generic 2kHz, or a chosen 12kHz wavelength. Any unsuspecting target will have difficulty 'timing', and then 'focussing' in on the location of the sound because the beeps will randomly vary in intervals, ranging from two to eight minutes. The default 2kHz sound is pretty annoying for adults, but if you really want to get a reaction, then select the 12kHz sound as it's the sweet-spot frequency—wise for annoying teenagers!

NOT WANTED

Kerala Trance Binary Watch. 'Hold On A Few Minutes, I Just Gotta Add This All Up . . . '

Kerala Trance Binary LED Watch
£60.99 (best price)

When a colleague (or stranger in the street for that matter) asks you for the time, you can forget casually glancing at your wrist and efficiently announcing the answer. This timepiece requires a lot more in the way of concentration and numeracy skills. Aesthetically it is very '80s retro-chic, and could comfortably grace any catwalk from Milan to New York, but on the functionality and ease-of-use front, it is about as simple as adding 32, 16, 8, 2 and 1. OK, so we admit that with a little practice you will be able to shave seconds off the calculation time, but then we're not all blessed with the mathematical skills of Rachel Riley!

NOT WANTED

Drive Alert. 'Wake Up!!' Don't Let Your Head Droop While Driving.

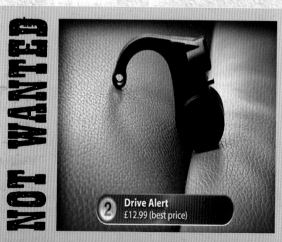

Drive Alert
£12.99 (best price)

Now, it's not the 'stopping drivers falling asleep at the wheel' idea behind Drive Alert we didn't see eye-to-chest with. In fact drivers falling asleep at the wheel is something that tech should help tackle. But it is the actual application and practicality of the device that put us off. To get it to activate Suzy Perry had to physically place her chin on her chest, which meant in the case of a driver actually falling asleep at the wheel that by the time it had woken the driver up, the air bag would be the next thing he or she would be confronted with, following its loud deployment bang that would have drowned out Drive Alert's supposed warning beep in any case.

NOT WANTED

Litter Kwitter. Forget The Tray And Share A Toilet With Your Feline Instead!

Now, not every cat's toilet routine is as domesticated and human-like as 'Jinxy' from the movie *Meet the Parents*, but with the 'Litter Kwitter' (as recommended by vets . . . allegedly) it is time to say goodbye to plastic trays and bags of litter and say hello to the purpose-designed cat-friendly 'over-seat' with its three-stage inserts construction. Thereafter, there will be the not so small matter of your cat hogging the toilet in the mornings while you are trying to get ready for work, and a cat flapped-adorned bathroom door. Apart from just getting them to stand on the seat long enough to do their business in the first place, would they flush adequately and wash their paws afterwards? Nice try, but we're firmly sticking with the traditional tray, scoop and bag method!

NOT WANTED

Puchi Puchi Electronic Bubble Wrap. Stressed, Need Some Therapeutic 'Pop' In Your Life?

If not, you probably will after buying one of these! The Puchi Puchi is available in a variety of bright and pastel colours and electronically (dare we say digitally) simulates that oh so satisfying feeling of popping the resistant bubbles of a sheet of bubble wrap. This gadget is annoyingly addictive, or addictively annoying (delete as appropriate) and will, among other things (like blisters and RSI) eventually reward you with a random surprise sound effect when you hit your personal popping century. Whatever happened to the old analogue sheet stuff? Ah, the good old days . . .

NOT WANTED

Das Keyboard. Not To Be Used In A Dark Room (Or A Lit One for That Matter!)

Perhaps learning to touch type or are you just plain masochistic? The Das Keyboard will probably make you want to give it 'Das Boot' after trying to use it for more than five minutes. We get that many professional typists don't need to see what is on the keys, and it is a measure of one's typing prowess not to constantly look at the keyboard, but isn't the best solution to simply either look up or away and use a normal keyboard that the rest of the family or colleagues can still make use of? If however you're into minimalistic design and the Henry Ford approach to colour options, then Das Keyboard is probably for you!

NOT WANTED

Road Pro 12-Volt Sandwich Maker. Stuck In A Traffic Jam? Hungry? Fancy Low Voltage Toasted Cheese?

Since the dawn of modern civilisation, it has been the dream of road users (and the odd student with access to a car battery) to have ready access to a 12-volt toasted sandwich maker. And now it is a reality. You never need be more than seven minutes away from the most exotic combination of cheese and 'additional' fillings you could ever want. Just a few words of warning: it gets very hot, cheese and chorizo stains are notoriously hard to get of car upholstery, and the inside of most cars don't make the best impromptu kitchens. I mean where do you put the chopping board or fit the sink? As a camping accessory it is passable, but steer well clear for in-car use . . .

3 Road Pro 12-Volt Sandwich Maker
£14.98 (best price)

NOT WANTED

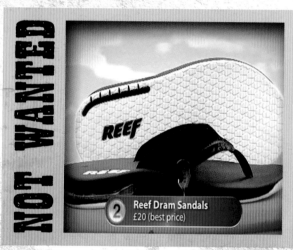

Reef Dram Sandals. Be The Heart And 'Sole' of the Party (Just Don't Tread In Anything On The Way!)

If we were still in the 1920s then this would have been a perfect 5 G-rated Prohibition gadget – an ingenious way of hiding illegal liquor from the authorities. But in these more liberal times the thought of first filling the sole of a sandal you plan on wearing all day with booze and then drinking from it pretty much turns the collective stomachs of *The Gadget Show*. And let's face it, even if you did take a wee podiatry dram, getting thrown out of a nightclub for hopping on one leg while drinking from your sandal isn't going to do anyone's street cred any favours – even if the sandals match your glasses and your name is Jason Bradbury!

2 Reef Dram Sandals
£20 (best price)

NOT WANTED

The Goatee Saver. Not Just The Facial Wear Of A Batman Villain, It Is The Shape Of Beards To Come!

Tired of trying to keep your goatee beard both even and symmetrically trimmed when shaving? Always taking too much off the sides and not enough from the bottom? Then the Goatee Saver is here to save the day. Just gently bite on the big red rubber mouthpiece (without gagging) and instantly transform yourself into both a character from *Batman Rises* and the most skilled barber outside Seville. Simply shave around the unit for that generic, style-free goatee beard that really exudes modern day 'uber-cool'. Either that or learn to shave properly, and shape one yourself that actually does look the part!

1 Goatee Saver
£13.39 (best price)

The Gadget Show's
Music Top Ten

'Mary had a little lamb...' may be a childish ditty, but on 12 August 1877, Thomas Edison used a tinfoil cylinder and a homemade stylus to both record and then actually playback these very words on the world's first 'reliable' recording device. Edison became known as 'The Wizard of Menlo Park' and was the founding father of all modern recording media, but that was only one of his many technological achievements.

He also developed the telegraph, telephone, phonograph, electric light bulb, alkaline batteries and 'Kinetograph' (a basic motion picture camera). Today's lightning fast data-transfer speeds and vast storage capacities all owe their existence to the perseverance and ingenuity of Thomas Edison, and it all began with a simple children's nursery rhyme.

Thomas Edison's Cylinder Phonograph

The Vinyl Album and 12" Single

Still the finest way to buy, keep and listen to music. Truly tactile and solid, vinyl has a character that's both collectible and precious. With both coloured and picture vinyl making certain releases stand out from the crowd and album art as collectable and varied as the contents it protects, purists and audiophiles still maintain that vinyl offers a warmth and depth of sound that even the best mastered CDs find hard to match.

In the guise of the 12" single and infamous 'White Label', the use of vinyl in combination with a pair of 'decks' and an audio mixer was the driving force behind the birth of the modern dance music scene and global DJ culture. Michael Jackson's 1982 album *Thriller* is still the best selling vinyl album of all time and New Order's *Blue Monday* the best-selling 12". Vinyl is having something of a resurgence of late, with both underground and major labels producing limited runs of certain albums and 'vinyl only' club nights re-asserting the power of the 12". With the feasibility of 3D printing albums that were never released on vinyl now a viable proposition, vinyl is very far from dead.

There cannot be a more iconic music gadget of the digital era than Apple's iPod. In the first decade since its 2001 release, Apple have sold in excess of 275 million units worldwide, and what started with a basic music only product with the marketing slogan, '1,000 songs in your pocket' has developed into the latest iPod Touch with its vast storage, Retina display, FaceTime video calling, HD video recording, Internet functionality and a dedicated Game Centre.

Not only have users created their own unique playlists from their personal music collections, they have downloaded over 10 billion songs from Apple's iTunes service since its launch! What started as just a simple digital music playback device is now a multi-media hand-held entertainment centre in your pocket, with up to 64GB of built-in storage and more processing power that many home computers of a decade ago.

The iPod

The late 1960s and early 1970s saw the widespread use of music compact cassettes. They quickly proved to be a much better option for car stereos than the far bulkier and relatively costly 8-track cassettes and players of the time. Masaru Ibuka, Sony's co-founder, apparently got fed up of lugging Sony's bulky TC-D5 cassette recorder around on business trips and asked his company to design and develop a compact playback-only stereo version, optimized for use with just headphones. On 1 July 1979 the company introduced the TPS-L2 to the World. Weighing in at just 14 ounces, the blue-and-silver portable cassette player had chunky buttons and two headphone jacks so that two people could listen to what was playing at once.

Originally the Walkman was introduced in America as the **Sony Sound-About** and in the UK as the **Sony Stowaway.** Eventually they sensibly landed on a universal name, the **Sony Walkman**. It was a huge consumer hit and upwards of 50,000 units were sold in the first two months in Japan. The tape based Walkman was discontinued in 2010 after selling over 220 million units worldwide, but Sony continue the Walkman legacy and still sell an MP3-based Walkman to this day.

Sony Walkman

Technics SL-1200 Turntable

If any music product could be rightly described as the industry standard, then it has to be the SL-1200 and -1210 turntables. Nicknamed by DJs, music producers and record collectors alike as 'The Wheels of Steel', they were manufactured by Matsushita of Japan. The original 1972 silver 1200 Mk1 models with their rotary pitch dials became 'decks' of choice for early scratching hip hop DJs. The S-1200 Mk2 was released in 1979 and introduced the now universally used slider pitch fader. It allowed the record's speed to be accurately and quickly adjusted by -8% to +8% from its original beats per minute – and the rest is music history.

Later Technics released the SL-1210 MK2 in a cool satin black metallic finish, and while the look was very different, the only real change was to the internal electronics to improve reliability. Last sold in their original form in 2010, the SL-1200s and -1210s are simply the best direct drive DJ turntables ever. With over three million sold, the Tec 12s have earned their iconic status and an example can even be found on display at the London Science Museum as part of their 'technology that shaped the world we live in' exhibition.

Noise-cancelling headphones are a must-have product in the world of personal audio, with many companies competing for the pound in your pocket. But the seeds of the technology were first planted when the late Doctor Amar G. Bose (MIT professor, entrepreneur and founder of the Bose Corporation) took a 1978 flight from the US to Europe. Using a set of airline-supplied headphones, he found that the sound quality was so badly impaired by the roar of engines it made listening to music virtually impossible. After many years of development, Bose released noise-cancelling headsets for a variety of applications, including use by both military and domestic pilots, combat vehicle crewmen, and more recently as 'on-ear' and even 'in-ear' variations for personal listening. Bose also used their noise-cancellation technology to develop headsets to protect the hearing of the crew of NASA's Space Shuttle and pilots undertaking the first un-refuelled, non-stop around-the-world flight in the 'Rutan Voyager'. Dr Bose we salute you for both protecting our hearing and improving our listening pleasure!

Bose Quiet Comfort Headphones

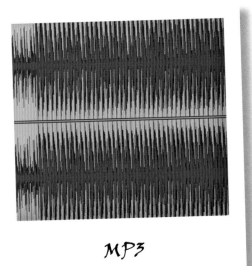

MP3

The compressed file format that changed music forever... MP3 stands for 'MPEG Audio Layer III' and it is a universally recognised standard for audio compression that makes any music file smaller with little or no loss of sound quality. MP3 is part of MPEG, an acronym for Motion Pictures Expert Group, a family of standards for displaying video and audio using slightly different compression codecs. The original research was done in Germany and led by Karlheinz Brandenburg, who is often called the 'father of MP3'. Brandenburg specialised in both mathematics and electronics and had been researching methods of compressing music without the associated loss in quality since 1977. What the resulting MP3 codec did was shrink down the original sound data from a 44.1 KHz recording at normal CD quality by a factor of 12. Amazingly this was achieved without any perceived loss in sound quality allowing far more tracks to be stored onto recording media. By 1997 the first commercially available MP3 player (the 'MPman') hit the market and the birth of the personal MP3 player had arrived...

With the advent of always-on Wifi and fast broadband connections, it was perhaps inevitable that dedicated music streaming services would emerge. Launched in 2008, Swedish Spotify kick-started the revolution of not actually 'owning' music. Instead, you can create custom playlists using the Spotify application and listen to virtually unlimited content. If however you do feel that you would like to own the track or album there is an option to pay and download it in a similar way to Apple's iTunes. The service currently has over 6 million paying subscribers, and 24 million active users with what the company call a 'Freemium' financial mode. Basic services are free but with a capped monthly listening allowance and advertising of up to 30 seconds in length placed between the tracks. More advanced features are offered at a premium subscription rate that also sees the advertising removed, and allows the subscriber to access their Spotify account on additional iOS, Android, Symbian, Windows Mobile, Windows Phone or BlackBerry devices. Criticised by some within the music industry as insufficiently rewarding artists, it is however a very cost-effective way of accessing an almost limitless resource of current and back catalogue music and is revolutionising the way millions consume music.

Spotify

Sing Star

The most successful, addictive and brilliant music game ever – and it's British too! SingStar undoubtedly cashed in on the popularity of karaoke, but used the technological power of both the second and third generation Sony PlayStation consoles to add another competitive edge to the proceedings. First released in 1994, and developed by London Studio, the SingStar games are sold either as the software alone, or bundled with a pair of USB microphones. Eventually wireless microphones were also made available allowing users a less restricted performance. The games are compatible with both the 'EyeToy' and PlayStation 'Eye' camera systems, which means players actually get to see themselves singing along on screen! Over thirty titles in the series have now been released; with 'Sing Along' tie-in editions such as 'Take That', 'Queen', 'ABBA' and even 'Disney', being very well received by the gaming public. Singstar is the perfect party game that the whole family can play, even the tone deaf ones!

The greatest electric guitar of all time and a more iconic example of what a guitar should both look and sound like there has never been. It has become the mainstay for artists like Slash, Eric Clapton and Mark Knopfler, and a must own item for any serious or aspirant guitar player. But who was the eponymous Les Paul? Well, Les Paul moved to Hollywood in 1943, where he played with Nat King Cole and other artists in the inaugural 'Jazz at the Philharmonic' concert in 1944. Paul built and used his own guitar that he affectionately called 'The Log'. It incorporated many technical innovations based around a solid body. Gibson Guitar Corporation designed a guitar using many of Paul's innovations in the early 1950s, and presented it to him to try. After his first session with the guitar he was so impressed he immediately signed a contract to endorse what became the 'Les Paul' model.

The arrangement persisted until 1961, when declining sales prompted Gibson to change the design without Paul's knowledge, creating a much thinner, lighter, and more aggressive-looking instrument with the now legendary two cutaway horns instead of one. Paul disliked this new model and asked Gibson to remove his name from the headstock. Gibson promptly renamed the guitar SG (which stands for Solid Guitar), and it would go on to became one of the company's best-selling designs. This aside, the original Gibson Les Paul guitar has remained a much sought after classic, and the design is still in production today. Paul eventually resumed his relationship with Gibson and endorsed the original Les Paul guitar design right up until his death in 2009. He will always be remembered as a true guitar hero.

Gibson Les Paul

Vintage tech ads

'The world's best.'

Published circa 1975

In the days before compact music cassettes, CDs or MP3 players, the reel-to-reel tape deck was king. Used by everyone from Hollywood sound technicians to musicians like Jimi Hendrix, the Beatles and Pink Floyd, the format allowed competent users to experiment with looping, splicing and over-dubbing. Sony's claim to be 'The World's Best' may have been a bold one, but at the time both audio technophiles and the recording industry hailed the Sony 880-2 as a technical masterpiece. With peak level meters, direct drive motors and a professional build quality, it was a sought after piece of audio tech. At ¥ 550,000 in 1975 it wasn't cheap … but then quality never is.

SONY 880-2
THE WORLD'S BEST.

DIRECT DRIVE SERVOCONTROL SYSTEM.

The name may be long—**Close-Loop Dual Capstan Tape Drive**—but the concept is simple: one capstan is just an extension of the motor shaft itself (the other connects through a belt-drive inertia fly-wheel). Gone are the intervening gears that can often impair optimum operating reliability as well as speed accuracy. The result—almost nonexistent wow and flutter—a mere 0.02% @ 15 ips.

PHASE COMPENSATOR CIRCUIT.

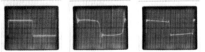

Original Source*   Source After Recording*   Source Through Phase Compensator*

Ideally, what you want on recorded tape is a "mirror image" of the original signal. No more. No less. Problem: the very nature of the recording process causes phase distortion. Solution: during playback, Sony's exclusive Phase Compensator Circuit compensates for phase distortion. Result: sound quality that's virtually identical to the original source. (REFER TO OSCILLOSCOPE READINGS.)

SYMPHASE RECORDING.

Thanks to the durability of Sony's **Ferrite and Ferrite Heads** and incredible precision fabrication and alignment of the head gap, you can record any matrix 4-channel signal (like SQ** or FM), play it back through a 4-channel decoder/amplifier, and retain the exact positioning of signal throughout the 360° 4-channel field. What started out in right front channel stays there. What began in left rear doesn't wander over to right rear. There's no phase shift whatsoever.

PEAK READING VU METERS.

They're versatile. Accurate. And incredibly informative. **1.** You can set for standard VU operation to determine recording level. **2.** Set to display transient peaks only (up to +15 dB). **3.** A third display, Peak Hold, retains transient reading, letting you accurately measure audio input and adjusts accordingly with 2dB **Stepped Record Level Attenuators.**

SYNCRO-TRAK.

This means you can lay down two individually recorded tracks in perfect synchronization with each other. Record head has playback-monitor function in record mode. This eliminates time lag that occurs when monitoring through playback head. Thus both tracks can be first generation, keeping noise levels at minimum. Flashing **Standby Signal** alerts you that the unrecorded channel is record-ready. And **Punch-In Record** puts you into record mode instantly, without stopping tape.

SONY. Ask anyone.

Brought to you by **SUPERSCOPE**

TC-880-2

TC-756-2

Other Distinguished Sony Decks   TC-788-4

*1000 Hz @ 0 dB, 15 ips. **TM CBS, Inc. (Side panels of these units are constructed of plywood, finished in genuine walnut veneer.) ©1976 Superscope, Inc. 20525 Nordhoff St., Chatsworth, CA 91311. Prices and models subject to change without notice. Consult the Yellow Pages for your nearest Superscope dealer.

Never Make Predictions, Especially About The Future

Baseball legend Charles Dillon 'Casey' Stengel

Ten Tech Predictions That Were So Wrong . . .

The Gadget Show doesn't really make a habit of making wild predictions about technology – things are moving way too fast for that! However, from time to time we do come across, and get to test, something or other that is destined to become the Next Big Thing. Often a gadget or a piece of new tech that will kick-start a huge lifestyle trend (Wi-Fi) or launch a host of me-too lookalikes (smart phones and tablets).

Notwithstanding our prophetic reluctance, there have been throughout science and gadget history a host of people who should have known better and refrained from putting their size thirteens in their collective mouths when it came to predicting the future of tech.

Often firmly and fatally convinced of their absolute rightness, it was only a matter of time before each were proved spectacularly wrong. Some were forced (literally in once case below) to eat their own words! A profound lack of imagination coupled with an unmitigated desire to maintain the status quo produces a mindset guaranteed to make an individual stand up and broadcast tech-forecast howlers, or, less egregiously, it is sometimes the case that the timing can be well and truly off – by a generation or more.

As *WIRED's* former editor-in-chief, Chris Anderson has said, 'It's interesting, when you look at the predictions made during the peak of the boom in the 1990s, about e-commerce, or Internet traffic, or broadband adoption, or Internet advertising, they were all right – they were just wrong in time.'

But then again some predictions have been just plain wrong-headed and so, with all due deference to *The Gadget Show's* collective Schadenfreude, here is a selection of our favourite tech forecast blunders.

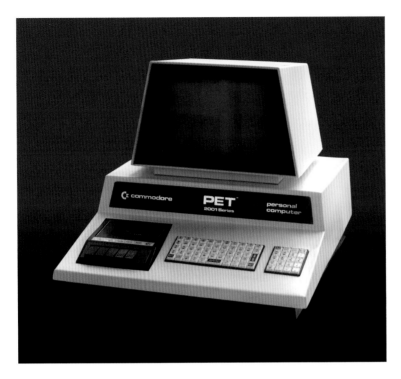

The Facts: In 2013 the number of personal computers worldwide surpassed one billion and, according to industry analysts, will probably exceed two billion by early 2014. The increasing global adoption of tablet computers has resulted in more mind-boggling numbers which in turn are dwarfed by the use of smart phones and the latest larger-screened phone-type devices nicknamed 'phablets'. All these very 'personal' computing devices are predicted to overtake traditional PC sales by the start of 2014. Apple's late and great Steve Jobs famously declared that we were entering a 'post-PC era' when the iPad was launched; less than five years later this is surely one prediction nobody could argue with.

Computers Aren't A Personal Device [1943 and 1977]

In 1943, the then head of IBM, Thomas Watson, Sr, was alleged to have said: *'I think there is a world market for maybe five computers.'* It should be noted at this point that despite repeated attempts to sift through the volumes of papers, speeches and reports Watson authored, no one has ever been able to find the original source of the quote.

So maybe we can withdraw the accusation of his utter wrongness and concentrate instead on Ken Oson, president, chairman and founder of Digital Equipment Corp who, nearly 35 years later, stated with gusto: *'There is no reason anyone would want a computer in their home.'* While a generation of modern-day parents may wish it were so, how much more wrong could he have been? The modern world wouldn't function without the personal computer; it is now permanently integrated into every aspect of life, from business to shopping, television and leisure time.

Would you want to live without one? We know we couldn't!

Apple Is Dead [1997]

Dell made his now infamous statement in 1997 when Apple was trading at just US$17 a share and struggling to get a real foothold in the computer market. Jump forward to October 2013 and the company is currently trading at US$482 a share and is a global brand of unassailable strength. Steve Balmer, the former head of Microsoft, along with Lord Alan Sugar, founder of Amstrad, both predicted the iPod would be a failure and, *'Wouldn't survive past Christmas . . . '* (perhaps Lord Alan should be fired from all prediction making after that little gem).

Today Apple has taken over the mantle as the world's biggest company and has more cash in the bank than most European countries.

In the mid-nineties an awful lot of people were busy predicting the imminent demise of Apple. Apple was, according to industry pundits, finished as a tech company, all washed up and ready for the corporate scrap heap. Michael Dell (founder of Dell Inc.) declared: *'Apple should shut down the company and sell the stock back to shareholders.'*

The Facts: 1997 was the year Steve Jobs replaced Gil Amelio at Apple, declaring himself 'interim CEO'. At the annual Macworld Expo that August, Jobs told the then relatively small band of Apple faithful that there was still hope for the company. Under his leadership, and through the marketing of iconic and visionary products, Apple has become the tech giant it is today. Steve Jobs died on 5 October 2011, and some argue that without his leadership Apple has started to lose its front-running status, but we are convinced that the legacy Steve created, and the passion for innovation and perfection he instilled into Apple employees, will ensure their long-term future and continued global success. Watch this space (preferably on an iPad).

9.7 Inches Is The Optimal Tablet Size [2010]

When the iPad was launched in 2010, Apple amazed the world with its startling vision of a new generation of (very) personal computing, to be known hereafter as the tablet. They claimed at the time that 9.7" was the perfect screen size for the new device, only to release the more compact 7.9" screened iPad Mini just two years later.

While the iconic and larger iPad is still seen as the 'serious' tablet for use in business or as a tech status symbol, its smaller sibling is the more convenient size and price for daily use in the educational sector, and for younger smaller-fingered users. Not so much a wrong prediction, more a creative adaptation to unveiled consumer needs. Still it's nice to know that even the tech-greatness that is the modern incarnation of Apple get things (slightly) wrong occasionally.

The Facts: Although Apple's 9.7" iPad's is undoubtedly the most recognisable tablet in the world, the market for more compact and portable devices grows daily. With sizes ranging from 5.8" up to 15" plus, it is a hungry market that many predict will overtake traditional PC sales within a matter of months. Interestingly, in the second quarter of 2013, 34.6 million Android tablets were sold compared with 14.6 million iPads – giving Android sixty-seven percent global share of the market. With models such as the Nexus 7 leading the charge against iOS and Windows devices, this figure is set to grow.

Cinema Is Just A Fad [1916]

You don't have to be an engineer, corporate leader or journalist to get things wrong (although it helps). The yet-to-be great comic actor of the silent era, Charlie Chaplin, rather remarkably claimed in 1916 that: *'Movies are a fad. Audiences really want to see live actors on a stage.'* At the time this assertion might well have had a ring of truth about it, but let's face it – he couldn't have been more wrong if he had taken a degree course in wilful misjudgement at the University of Wrongness. While theatre has a continuing and strong constituency, cinema-going is a dominant form of entertainment.

Admittedly, the advent of home cinema has caused some decline in the numbers, but box office receipts in the US still top US$10 billion annually. With giant cine complexes, digital surround sound, IMAX, 3D and with films such as *The Hobbit* shot in High Frame Rate High Definition, cinema is no longer just about watching a film. It is a visceral all-subsuming sensory experience. Something we can hardly blame Charlie Chaplin for failing to foresee.

However, take the time to watch his greatest on-screen work and you quickly realise that expensive film effects showcased in a tech-laden cinema is by no means the only guarantor of a great movie experience. He was right about one thing – audiences want to see real people doing and experiencing real things with which they can empathise, even if they are up there on a silver screen.

The Facts: Between 1914 and 1967 Charlie Chaplin wrote the scripts for, and starred in, over eighty movies and was credited as director of over seventy. He is still considered to be one of the most important stars of the early days of Hollywood. Audiences packed cinemas night after night to see his sublime *The Kid* and *The Gold Rush*. In 1932 it was estimated he was worth £1.5 million, a huge sum for the time, so, no doubt, he was relieved that his rather negative 1916 prediction about the future of cinema proved so wide of the mark.

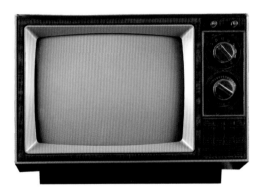

Television Is A flash In The Pan [1927]

Echoing Chaplin's doubts about the future of the cinema, the movie industry would later have its own suspicions about the viability of television when it first came knocking. Maybe it was an attempt to head off the threat to the film industry, but Twentieth Century Fox co-founder Daryl Zanuck (b. Sept 5 1902 d. Dec 22 1979) claimed that TV wasn't going to stick around for long, saying in a phrase of staggering pig-headedness: *'People will soon get tired of staring at a plywood box every night.'*

Love it or hate it, television has been a large part of everyday life for decades now. With all due respect to radio (and latterly the Internet), no other device has brought humanity's seminal life-affirming, life-changing or profoundly disturbing moments into the homes of our planet's citizens in such numbers and with such undoubted potency.

From our greatest achievements and most relished entertainments – the triumphs of the first moon landings, the fall of the Berlin Wall, the soccer World Cup; to the up close and visceral horror of war; and the defining moments of our shared history – the assassination of John Kennedy and the tragedy

The Facts: It is estimated that there are in excess of 1.5 billion television sets in the world, with China and America leading the way with over six hundred million sets between them. In the United Kingdom we have a modest thirty-five million sets in current use. It is estimated that television viewing takes up, on average, three to four hours per day in a typical household. Things are changing though, and with the increased penetration of online streaming of live programs and 'on demand ', 'iPlayer' and 'watch again' services, all available on computers, tablets, smart phones and increasingly via games consoles, the traditional television, with its attendant antenna, cable or satellite box, is no longer the fixed point around which families gather. But we're no Daryl Zanuk, and are a long way from predicting the death of the goggle box.

of 9/11, no other communications medium has othe reach and power to inform and entertain. Regardless of where you live, the language you speak or the profession you occupy, most people find themselves in front of a television at some point during the day.

Big Passenger Planes Will Never Happen [1933]

Shortly after the inaugural flight of Boeing's 247, 10-seater plane in 1933, a company engineer proclaimed (and you have to think he had an annoyingly self-satisfied look of entirely misplaced smugness on his face): '*There would never be a bigger plane built.*' In fairness, Boeing had a lot to crow about – the 247 was the first aircraft we would recognise as a modern passenger-carrying airliner with its all-metal construction and retractable landing gear, capable of crossing the United States from east to west eight hours faster than its predecessors.

But, let us remind ourselves, it carried a mere ten passengers and therein lay, notwithstanding the purblind boasts of Boeing, its certain demise. Fast forward to the modern day and we have, of course, a new world of double-decked jets that can carry up to 853 passengers and crew, with a colossal four-engine, eighty metre wingspan.

The Facts: The 247 had a top speed of 322km/h and a range of about 1,200km. The Airbus A380, on the other hand, has a top speed of 1,020km/h, a cruising speed of Mach 0.85 (eighty-five percent of the speed of sound) and range of 15,700 kilometres (capable of flying from New York direct to Hong Kong). Boeing themselves have also developed a long-range wide-bodied jet, the 787-10 Dreamliner. It is touted as a 'super-efficient airplane' using twenty percent less fuel than any other plane of its size, but the model ran into technical difficulties during its first year of service, with some airlines temporarily grounding it. And when you consider the Dreamliner 'only' carries around 330 passengers, you can understand why Airbus is leading the way in the really big plane market!

The Internet Will Fail [1996]

Bob Metcalfe, founder of 3Com Corporation and inventor of Ethernet (the cable that links most physical networks), famously predicted in a column for *InfoWorld* that the Internet would: *'Go spectacularly supernova, and in 1996 catastrophically collapse.'* When that prediction manifestly failed to come about, Metcalfe engaged in a gloriously public demonstration of penitence.

In front of an invited audience he placed the now discredited column into a blender, poured water over it and proceeded to eat the resulting 'word-smoothie' with a spoon. At least the guy had a sense of humour!

The Facts: With an estimated 2.4 billion users globally, the Internet is now almost universally available (in the Western world at least) and has an influence on most aspects of modern life – our shopping habits, the way we socialise, play games or watch films and television. It is thought that a considerable number of people are spending upwards of twenty-four hours a week online, and the hourly online figure is rising. By 2015 it is anticipated that over 960 exabytes of data (an exabyte equals one billion gigabytes or one quintillion bytes) will be consumed every twelve months. That is enough data to store nearly ten million years of 1080p, Blu-ray HD footage or stack data CDs from the earth to far beyond the moon. With 'all you can eat' 3G and 4G mobile data packages and unlimited home broadband tariffs becoming relatively affordable, having access to the Internet 24/7 is now the 'norm' rather than the exception.

Apple Will Never Bring Out A Phone [2006]

New York Times columnist, David Pogue wrote in 2006: '*Everyone is always asking me when Apple will come out with a cell phone. My answer is, "Probably never."*' Just one year later the first iPhone appeared (that must have really sucked!).

By the end of the second quarter of 2013, Apple had sold over 350 million of the devices around the world. With the launch of the aluminium-chassis iPhone 5s and lower cost plastic bodied 5c models breaking records by selling nine million units in just three days, Apple proved, if anyone doubted it, that they were still a massive player in the smart phone market.

With hindsight, under Steve Jobs' innovative uncompromising management and design ethos, Apple not only 'came out' with a phone, but also in doing so, revolutionised the whole idea of what a mobile phone should be, and helped redefine its place in our lives.

The Facts: iOS-based smart phones are actually the second best selling devices. Android powered models, such as the Galaxy S4, continue to dominate the market, with Windows-based phones gradually gaining traction. Apple's intentions with the lower-cost model are clear. It wants to attract the type of overseas and domestic customer who opted not to pay the premium price tag hitherto associated with previous incarnations of the iPhone. Apple is seeing huge sales growth in Russia, India and particularly China, where it is currently losing out to lower-priced domestically manufactured models – powered by Google's Android operating system.

Music Streaming Has A Bankrupt Business Model [2003]

Steve Jobs was right about many things, and his mould-breaking iTunes digital music sales model was the most exciting development in media distribution for a generation.

However, when he said that streaming music was 'bankrupt', he may have been somewhat off the mark. There are a number of music streaming ventures out there, but one of the best known and perhaps the user-friendliest service is Spotify. Reportedly, it has more than six million monthly paying subscribers and a further eighteen million users of the free advertising-driven service, all with access to its twenty million song catalogue.

Despite complaints by a number of artists that the royalty rates paid on streaming tracks is miserly, it seems that streaming is here to stay. With Nielsen SoundScan reporting the lowest weekly total of US album sales in July 2013 since the service began tracking sales in

1991, music streaming services are set to be either the salvation or the damnation of the music industry. The jury is out.

The Facts: 'Why would anyone want to purchase a record ever again if you can just listen to it for free?' It is the question most musicians ask when quizzed about on-demand streaming. Insufficient rewarding of musicians through an adequate royalty system – and thus helping to ensure the future of new artists – has been a criticism levelled at Spotify, but it pays out a figure approaching seventy percent of its revenue generated from advertising and subscription fees directly to the rights holders: artists, labels, publishers, and performing rights societies. With other streaming services such as the new free offering from Apple – iTunesRadio, Amazon's Cloud Drive and Xbox Music, it is clear that streaming is going to be an increasingly important way in which music and movies are distributed to the consumer – even if the business model remains ambiguous at best.

Spam Will Be A Thing Of The Past [2004]

Bill Gates seems to like his tech predictions: on this occasion claiming, in 2004, that spam would be eradicated by 2006. Unfortunately, Bill got this one wrong! Recent studies have shown that spam still makes up an estimated eighty percent of all emails landing in our inboxes or at least those that have forgone the protection of either active filtering or targeted spam blocking software.

Requests from 'Frank in Lagos' for our address and bank details so he can discuss with you how: *'One hundred and fifty million US dollars is able to transfer to you direct.'* are still on the rise. Whether someone chooses to act on such unsolicited emails is a matter of judgement, but any email that starts: **URGENT MESSAGE** or **BECOME MORE ATTRACTIVE TO THE OPPOSITE SEX** usually isn't and definitely won't!

The Facts: The art of spamming has become increasingly sophisticated, regularly using heightened levels of public interest in the biggest news stories to target the gullible. Attackers send out emails with supposed links to breaking news, but in fact on clicking the links readers are directed to a host of services and products. Time wasting, but not necessarily damaging. The links can be malicious though, and it pays to use one of the many free or paid for spam filter programs available, most of which work pretty effectively.

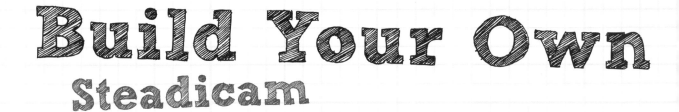

Build Your Own
Steadicam

Jason sets himself the task of building a DIY budget camera stabilizer for only £50!

In the mid-1970s, renowned cinematographer Garret Brown invented the Steadicam and in so doing revolutionized the film industry. His invention mechanically isolated the movie camera from the operator's movements, even when moving quickly or over uneven ground. Today it is used widely in sports broadcasting, documentaries and drama productions and gives a smooth shot without the shaking and jostling that would occur if the camera was merely hand held.

All camera stabilizers work on the same theory: you put your camera on the top of a vertical tube and on the bottom you add some kind of weight. If you get the positioning right, you create a counter-balancing effect (similar to the pendulum of a clock).

List of Build Items
Camcorder (roughly £500)

Components used:
Monopod (£16 each)
Jubilee clip
Corner Brace
Horizontal arm (£9.99)

Jason's own camera stabilizer took a few weeks to get right. He started out by using various designs with parts like copper piping, sundry mountain bike parts, various bits of plumbing equipment and fisherman's lead. However all this became too cumbersome and it occurred to him that off-the-shelf parts were going to be his best bet.

PLUMBING BITS AND PIECES AND BIKE PARTS. THE FIRST PROTOTYPE IS EVENTUALLY ABANDONED

THIS IS GOING TO WORK!

A JUBILEE CLIP FROM A LOCAL DIY STORE JOINS TWO MONOPODS.

So, Jason shows us his first bit of tech – a monopod. It's like a tripod but only has one leg that creates the basic steadying effect. Of course a camera monopod will have a typical camera mount already attached to the top of it (usually called a 'head') – so it is ready to hold your camera or camcorder.

ELEMENT NUMBER THREE IS A HORIZONTAL ARM FITTED TO THE BOTTOM OF THE VERTICAL MONOPOD.

THE HORIZONTAL ARM COULD BE
USED AS A TABLE CLAMP.

Jason then adds another monopod as the handle of the steadicam. This second monopod has a ball and socket joint at the top allowing Jason to use the handle to stabilize the camera mounted on the first (and vertical) monopod. Jason adds the handle as close as possible to the centre of gravity, where the stabilizing effect will be at its most pronounced, using a jubilee clip and a corner brace bought from a local DIY shop.

To stabilize the system even further Jason adds element number three – a horizontal arm that fits to the bottom of the vertical monopod. Jason will attach an added weight

A COUNTER-BALANCING WEIGHT WILL BE
ADDED TO THE HORIZONTAL ARM

THE WEIGHT IS AFFIXED. YOU COULD USE
ANOTHER CAMERA FOR A SECOND FEED!

to the far end of this. Professional cinematographers sometimes use this far end of the lower arm to clamp on to a table or to affix an additional light.

The final piece of this build requires a weight and Jason uses an old Casio video recorder, but you could use anything for weight such as lead. With the recorder attached Jason in theory could hook this up to his camcorder for an additional video feed.

Jason finally gets to test his steadicam. When compared with a full-blown professional steadicam kit, that at the time would have cost as much as £20,000, his relatively cheap £50 stabilizer works incredibly well.

WHAT DO YOU KNOW ABOUT?
Jon Bentley

Jon's life has been dominated by two passions: cars and technology. The addiction to cars led to his first job at the European headquarters of the Ford Motor Company, and then in 1984 to BBC's *Top Gear*. As Executive Producer of the show he was responsible for launching the television careers of Jeremy Clarkson, Quentin Willson and Vicki Butler-Henderson.

In 2002 Jon launched *Fifth Gear* for Channel Five, producing the show for two years before taking his rightful place in front of the cameras! When *The Gadget Show* was launched Jon joined the presenting team and has subsequently appeared in sixteen series of the show.

Here are five things you might not know about Jon . . .

1   I built my first (working) radio when I was seven and had a business mending and selling televisions at school.

2   I took my A-levels when I was sixteen. Later I went up to Oriel College at Oxford – the same college Rachel would study at!

3   I have about seven thousand car brochures. I stopped collecting them in the nineties but can't bear to get rid of them.

4   I did have my IQ measured as a child (I think it was fashionable at the time) and it was 164. However, I have never regarded IQ figures as particularly important.

5   I still own my first car, a VW Beetle.

A day in the life of
The Gadget Show

Director David Leighton runs through the script once more with Rachel and Jason.

So you call yourself a Gadget Show fan? Watch the show every week? No doubt you marvel at the in-depth knowledge of the presenters, love the wit and banter in the studio and relish the sheer energy with which the team goes about testing all manner of gadgets to destruction. But what exactly goes into making *The Gadget Show* the finest tech spectacle on the planet?

"Are we ready? Action!" Scripts are quickly hidden behind the sofa, the new Gadget Show studio goes quiet as Jason and Rachel immediately launch into a high-octane perfectly synchronized introduction to yet another Gadget Show test (this time it's cordless vacuum cleaners).

"Let's ago again," says Jason, "I can do it much better." David Leighton – long-time director of the show – emerges from behind the set where two small monitors are located (one for each camera) and agrees. While the shot is set up again, David takes the opportunity to ask Rachel to move a little closer to Jason and suggests the pair grab a handful of the vacuum cleaners and run out of shot when they have delivered the twenty second script. Cameras are repositioned and it is take two. Jason is right; this time the delivery is sharper, has more energy and the banter between Rachel and Jason has everyone laughing. Another episode of the world's best tech show is well on its way to our screens.

Filming Jason grabbing a handful of vacuum cleaners and running off with them turns out to be much trickier than anyone thought. Take three (or is it four?).

Somehow Rachel manages to make the task of spreading household dust look like fun.

The set up. Cameramen Mark and Chris arrive at the Birmingham Gadget Show HQ with mounds of kit and begin the process of setting up the studio for the day's shoot. The production office that sits next door to the studio is rapidly filling up with members of the research and production team. Jason is munching on a sandwich talking to everyone who will listen while the sound of a loud hairdryer means Rachel is already ensconced next-door with Sam, the show's regular make-up artist.

9.00AM

Ready and waiting. Director David Leighton has arrived and is chatting to Jason who, as usual, is a font of ideas about today's planned tests. Researchers Tadhg Leonard and Andrew Parkes are busy checking and rechecking the gadgets. Particular attention is being paid to topping up batteries as this is going to be a cordless test and battery life will be a key measure of performance. Jason and Rachel are miked up by sound engineer Simon as this morning's scripts are handed out.

10.15AM

In the studio. It is all remarkably informal, at least for now. David and the crew head into the studio and climb the stairs to the new mezzanine balcony now familiar to the show's fans. Cameraman Mark has already made sure the lighting is set up correctly and the sofas (there are two) are out of shot. They will be filming Jason and Rachel with their backs to the balcony railing with an array of cordless vacuum cleaners in front of them. Jason bounds up the stairs and Rachel follows a few moments later quietly rehearsing her lines. Suddenly we're down to business. David explains very clearly to the entire team exactly what he wants to achieve and, as the presenters quickly position themselves, the cameras are located for the optimum shot. We're ready for the first take of the day.

10.35AM

The test (part one). Filming Jason and Rachel 'running out of shot' proves more technically challenging than you would think. It takes several attempts with cameras sometimes following the action, sometime remaining fixed on the spot where the presenters had been. Eventually David is happy; he has enough footage in the can (or more accurately high definition video on SD cards) so that the show's production editor will be able to craft it into a four second segment later on.

10.45AM

The test (part two). The day before the studio carpet had been marked up with lanes. Now the seven cordless vacuum cleaners stand at the head of each lane like so many one hundred metre sprinters, while bowls of household 'dirt' that includes salt, dust, ash and cat hairs (which Jason will prove to be highly allergic to!) are ready to be emptied over the carpet. Jason and Rachel enjoy themselves immensely tipping the contents of the bowls over the carpet.

10.55AM

The test (cont'd). The next forty-five minutes is all about a rigorous, unforgiving and genuine test of the candidate cleaners. Moving from one to another and down the length of the carpet

There is nothing staged about Gadget Show tech tests and Rachel and Jason compare notes endlessly as filming progresses.

with its different bands of household dirt, Jason and Rachel, unprompted and unscripted, voice their opinions as they put each cleaner through its paces. Some fare well, others less so. In fact Jason abandons one cleaner in frustration early

in the test. It will lie forlornly on the carpet for the rest of the shoot. Only a small percentage of the presenters' ad-libbed comments will actually make it into the show; the test segment will eventually have a scripted voice-over added

An example of just how seriously *The Gadget Show* takes its testing: the crew use a borescope (an optical device usually used in the visual inspection of aircraft engines!) to scrutinize the amount of dust left by each vacuum on the hotel carpet.

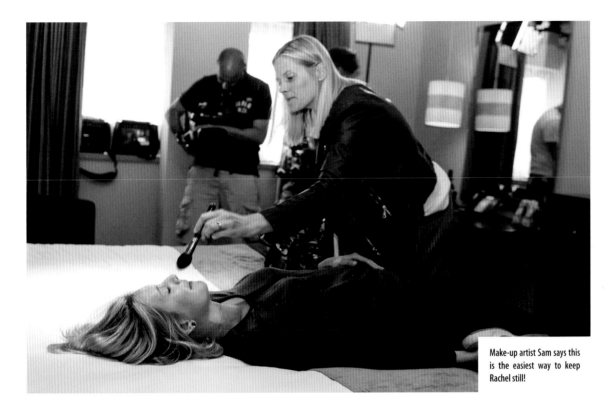

Make-up artist Sam says this is the easiest way to keep Rachel still!

Love the hair, Rachel!

once the full results of the test have been analyzed by David and the production team.

11.40AM

Outside filming. With the studio test in the can, it's time for everyone to decamp to a local Birmingham hotel where the top three cleaners so far will be tested in a real-life environment. Several vehicles draw up outside *The Gadget Show* offices and are loaded with piles of camera, sound and lighting kit. At the hotel two bedrooms have been reserved, and are soon being transformed with lights and cameras into mini studios to the mild bemusement of hotel staff. An hour later

and the second round of tests is done. The consensus is that there is a clear winner, but that all of the cleaners suffer from the problem of frustratingly short battery life.

13.00PM

Lunch. Everyone grabs a quick bite at a local café apart from Rachel and Sam who head back to the production office. Rachel has to have her hair platted for the afternoon filming!

14.00PM

Costume change. Jason and Rachel are required to change into camouflage t-shirts and trousers as the plan is to give this afternoon's destructive test of rugged outdoor Bluetooth speakers a military-style look and feel.

14.30PM

On the road. It's back into the cars and a drive to an assault course that has a twenty-metre tower complete with a zip wire. Ever-energetic Jason has us all laughing with stories of his long-forgotten television show that sounds like an extreme version of *Beadle's About*.

15.00PM

The afternoon test. These days it is hard to imagine going camping, sitting in the park or garden on a sunny afternoon or relaxing by the pool without our favourite tunes playing in the background. But we're talking about Great Britain and our wonderfully unpredictable weather. A battery-powered device exposed to the elements is not necessarily a good idea, but a number of manufacturers have developed Bluetooth-equipped speakers with portability and for use in electronics-adverse conditions in mind. Jason and Rachel are all set to give nine of these devices a severe military-style check out.

'Rugged' is a less than precise term, but the industry has developed standards based around Ingress Protection (IP,) and an IP rating indicates how well a product's casing resists solid particles like sand and dirt and liquids (especially water).

Somehow though I don't think they ever intended what *The Gadget Show* team have in mind.

'Ready for my close up, Mr. Leighton' purrs Rachel. Jason is just enjoying the moment.

The water test. It's a bright sunny afternoon, perfect summer weather with no prospect of rain. So with cameras set and Simon happy with the sound, Jason and Rachel begin by attaching chains to each of the speakers (mostly with duct tape). Then one by one, with music being streamed from an iPad, they are dunked like so many sixteenth century witches into the local canal. Jason is adamant, "If they don't come

out of the canal sounding as good as when they went in, they don't make it through to the next round!"

Quite a lot don't.

And the next round consists of Rachel and Jason running to the tower (about thirty metres away and about which more later) twirling each speaker on its chain, smashing it into trees

and other obstacles as they go. There are two objectives – (1) to see how well the speakers stand up to (considerable) physical violence and (2) how well they maintain the Bluetooth connection over distance.

16.30PM

The tower test. Surprisingly all of the remaining speakers are still in one piece and the destructive antics so far haven't whittled

Jason takes no prisoners. The tests can be truly destructive.

What can be more fun than ducking expensive tech in a canal!

'Not sure this one has survived, Rachel!'

the numbers down sufficiently. David calls a halt and throws it open for discussion. In a remarkably democratic atmosphere anyone can make a suggestion and eventually a production assistant, whose sole function so far has been to hold a GoPro camera over the canal on a stick, proposes that each speaker is attached to a rope at the bottom of the tower by Rachel and is then hauled up to the top by Jason. With the streaming iPad at the bottom of the tower,

Rachel does her piece to camera before hooking up another ill-fated Bluetooth speaker. Chris has avoided an afternoon of running up and down the tower.

surely some will lose connection before the top? The height at which they go quiet can be logged, thus giving us the top two finalists. Genius!

All this comes as a huge relief to cameraman Chris who was originally scheduled to run up and down the tower filming Rachel and Jason walking each speaker to the top one by one.

16.45PM

The tower test cont'd. Eventually enough rope is found in a locker and half the crew, along with Jason and David, climb the twenty-metre plus tower. Cameraman Mark, who is utterly unfazed by heights, is in the hands of the crew safety experts as he is strapped into a harness and hooked on to a rail. He steps over the edge and starts filming Jason hauling on the rope.

It's a slow methodical process as each speaker is dragged to the top and the point at which the connection to the iPad fails is carefully logged.

18.00PM

The zip wire. And we are finally down to two. All afternoon Jason has been eagerly eyeing the zip wire. "What if I go down the wire with one speaker and Rachel follows with the other and we work out which one got the furthest before we lose the signal?" Brilliant idea agrees David and starts to organize how it will be shot.

18.20PM

The zip wire (cont'd). With a carabiner, the Marmitek BoomBoom 260 speaker is hooked to Jason's belt and with final instructions from the safety officer he clears the tower. Two cameras track his rapid descent.

A few moments later Rachel appears at the top of the stairs and is quickly fitted with a harness before being attached to the zip wire. In reality she is loving every moment of it, but plays to the camera pretending to be scared. Away she goes and that's it – filming is done for the day.

19.00PM

The kit is packed away, the research team makes sure the tech has been logged and is back in the

Jason ready for the zip wire.

Rachel, harnessed and GoPro'd up, is not so sure it is such a good idea after all.

car (it will all be returned to the manufacturers) and the scrupulous notes of the tests are carefully cross-checked and filed.

We head round to the site manager's office to thank him and his team for providing such terrific facilities and their willingness to help whenever called upon to do so.

19.30PM

Back at the office. The cars are unloaded and the footage is handed over to the edit suite next door so that it can be immediately backed up. Tomorrow it will be reviewed by David and the production editor who will spend a day selecting which shots will make it into the final edit.

20.00PM

Everyone is about ready to call it a day. Jason and Rachel leave for their local hotel (they are both due back in the studio at 8.00am in the morning) while Mark and Chris begin sorting the cameras they will need tomorrow. Two terrific segments are safely in the can. And while it's been a long day, everyone is really pleased. All in all, another hugely productive day in the busy life of the UK's favourite tech show.

The day's footage is already in the editing suite.

Vintage tech ads

'You have to see it to believe it!'

Published circa 1955

Today, using a television remote control is second nature, but in the 1950s when you wanted to change channels you got up from your vinyl-covered sofa and pushed buttons or twiddled knobs on the actual television itself. How quaint! Eugene Polley's Flash-Matic remote control was therefore, something of a revelation. The sci-fi looking ray-gun shot a beam of light at receptors located at each corner of the screen that would apparently change channels, turn on and off the picture and even alter the volume.The dawn of the couch potato was upon us!

YOU HAVE TO SEE IT

FLASH-

With a beam of magic light

this Zenith "flash tuner"

works TV miracles!

Absolutely harmless to humans!

W H

Here
devel
Just t
easy c
Flash-
nels.
comm

YO

BELIEVE IT!

MATIC TUNING

BY ZENITH

ONLY ZENITH HAS IT!

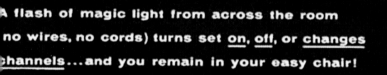

A flash of magic light from across the room (no wires, no cords) turns set **on**, **off**, or **changes channels**...and you remain in your easy chair!

CAN ALSO <u>SHUT</u> <u>OFF</u> LONG, ANNOYING COMMERCIALS PICTURE REMAINS ON SCREEN!

The Bismarck (Model X2264EQ), 21", Flash-Matic Tuning, Cinébeam®, Ciné-Lens. Blond grained finish cabinet on casters. Also in mahogany color (X2264RQ). As low as $399.95.*

amazing new television and only Zenith has it! without budging from your can turn your new Zenith t *on, off,* or *change chan-* even *shut off annoying* ile the picture remains

on the screen. Just a flash of light does it. There are no wires or cords. This is not an accessory. It is a built-in part of several new 1956 Zenith television receivers.

Stop at your Zenith dealer's soon. Zenith-quality television begins as low as $149.95.*

If it's new...it's from Zenith!

VE TO SEE IT TO BELIEVE IT

Manufacturer's suggested retail price. Slightly higher in Far West and South.

ZENITH®

The royalty of TELEVISION and radio
Backed by 36 years of leadership in radionics exclusively
ALSO MAKERS OF FINE HEARING AIDS
Zenith Radio Corporation, Chicago 39, Ill.

The Gadget Show's
Camera Top Ten

Invented by Frank A Brownell and costing just one US dollar when launched, the first 'Brownie' camera came with a cardboard shell and was capable of taking six exposures without re-loading, producing a square picture size of 2 ¼ x 2 ¼ inches. In one year over 150,000 examples were shipped. Named after the popular 'Brownie' characters of the 1880s created by author and illustrator Palmer Cox, it was the camera that anyone could actually own.

The Brownie saw the start of the widespread adoption of photography as a medium and hobby for everyone, not just professionals with their cumbersome equipment. The design evolved into numerous variations – the most popular being the 127 model that had three incarnations and sold millions of units between its launch in 1952 and the end of its production in 1967.

*Kodak Brownie
(From circa 1900)*

Leica I Model A (1925)

Launched in 1925 at the Leipzig Spring Fair, the Leica I Model A was the camera that really put 35mm photography on the map. It was the first high-quality 35 mm cameras to be mass produced. From here on, the design defined the direction photography in the 20th century would take, and shaped the basic look and layout of most 35mm cameras. More importantly it established the viability of 35mm as the format of choice for both professional and amateur photographers alike.

The model A has also been noted as the camera that made the Leica brand universally known as a manufacturer at the forefront of camera technology. One often overlooked fact is that Leica supplied an especially modified monocular for the NASA Apollo 11 mission, which became the first optical device ever used on the moon.

Polaroid Land Camera Model 95 (1948)

A true innovation in camera tech, the Polaroid Model 95 was the first 'instant' camera that allowed users to see the result of their photos after a wait of just a minute. The process, invented by Polaroid's founder Edwin Land, used a diffusion transfer to move the dyes from the negative to the positive via a re-agent. By 1956 Polaroid had sold over a million cameras worldwide, and by 1970 total sales topped US$500 million.

After a much-publicized lawsuit and a near 15-year battle with Kodak (for the latter's copyright infringement), Polaroid eventually received a settlement in excess of US$900 million. Although the giant has now lost its grip on the camera market, the term 'Polaroid' lives on, and the company itself is still innovating today with their latest generation of instant and digital cameras, camcorders, printers, tablets and audio devices.

With the 'F' taken from the letter F in 'Reflex', this early Nikon camera was the first truly professional-calibre SLR. It dominated the professional market long before brands like Canon came along and offered real flexibility with its array of add-ons. In 1959 this really was cutting edge of camera tech and the company has been building on this success ever since. Although many of its built-in technical concepts had already been introduced elsewhere, the F was deemed truly revolutionary as it was the first modern camera design to combine them all into one single unit.

From its 1959 launch, it was produced right through until October 1973, eventually being replaced by the very similar Nikon F2. Aspects of its original design remain in all of Nikon's subsequent SLR cameras including the current Nikon film and digital models. Historically Nikon F cameras were used to produce some of the most striking images of conflict in places like Vietnam and Korea, and to record the launches of the Mercury, Gemini and Apollo space missions.

Nikon F (1959)

The DC 210 was the world's first 'megapixel resolution' digital camera selling for less than one thousand dollars ($899), which for 1998 was a real breakthrough. The basic specs included a removable 4mb CF card, a 2x wide angle zoom, high-res CCD sensor, TV/video output, built-in infrared communications, and 'finished file format' processing in the camera itself. This allowed the CF card to be plugged directly into a card reader and files to be manipulated immediately on a computer. Powered by just four AA batteries, it wasn't exactly pretty, but was the first popular digital camera that combined point-and-shoot convenience with many features we now see as standard on modern digital cameras.

The megapixel design meant the camera achieved a quality level that, at the time, matched conventional point-and-shoot film cameras, at least up to the 6 x 4 inch print size most commonly found in consumer photo printing. In a marketplace we take for granted today, the DC210 led the way in the digital camera revolution.

Kodak DC 210 (1998)

In November 2000 the Sharp H-SH04 'J-phone' was released in Japan and has been universally recognized as being the world's first 'integrated' camera phone. With its futuristic looks and ease of functionality, it paved the way for the hundreds of camera phones on the market today and its near universality as the most-owned piece of tech in the world. Combining the two products – phone and camera – gave the user the chance to not only stay in touch, but to always be able to capture that special moment for posterity.

The resolution may not have been ground breaking with a 0.11 mega pixel CMOS and a built-in 256 color display, but it was the moment and the concept that changed the way we use our mobile phones, take pictures and eventually share them via social media today.

Sharp H-SH04 'J-Phone' (2000)

Canon 5D MKII (2008)

The camera that transformed DSLRs from being a device solely used for the production of still photographs to one capable of capturing incredible HD video. It actively led the way in creating a new style in the video sector and was also the basis of the widely used broadcast mainstay the C-300. With an impressive 21.1 megapixel, full frame CMOS sensor, and an ISO range expandable right up to 25,600, full HD 1080 at 30fps movie recording and a built in hi-res 3.0 inch VGA LCD with Live View, the Canon 5D MKII isn't just an amazing camera capable of stunning location or studio stills, it has also become the mainstay of both amateur and professional filmmakers alike.

Hit TV shows like *House* have filmed entire episodes using the 5D, and Canon have confirmed that 5 MKIIs were used when filming point-of-view action shots in the recent Marvel 'Avengers Assemble' movie that were then be seamlessly intercut in the edit with footage from the film's principal 35mm and digital cinematography cameras.

Maybe not one of the most iconic cameras, but we absolutely love it on *The Gadget Show* and it was a top camera tech pick by Jon Bentley for many years. The feature list is impressive for such a compact camera with a 12x zoom and image-stabilized Leica lens plus a focal length range equivalent to 25-300mm. Other features include an ultra-sharp 3.0 inch 460,000 dot wide-view LCD monitor, a dual-processor and most significantly a 1280 x 720 pixel HD video recording with stereo sound and full optical zoom capability.

It also surprisingly uses the advanced 'AVCHD Lite' recording system, a video encoding format designed for solid-state camcorders. The TZ range still exists and continues to develop – as does the massive zoom inside its compact shell.

Panasonic DMC TZ7 (2009)

The latest iteration of the famous GoPro brand that has transformed the way action-based movies are shot. Not only has it been embraced by the extreme sports world and amateur filmmakers, but is widely used by professionals too. It is regularly used by broadcasters including nature filmmakers, and we'd never be without a battery of GoPros right here on *The Gadget Show*. This latest HERO3 model offers wireless control, real-time video streaming, a wider range of resolutions and filming modes, and takes its design to a whole new level of shock- proofing, waterproofing, and even (by linking two cameras together with a special lead in a double enclosure) the possibility of recording in three different formats of 3D.

Others have tried to take GoPro's crown, but none have offered the ease of use, the range of accessories and professional quality mounts, or – the bottom line – the downright robust dependability that the HERO3 offers.

GoPro HERO3 (2012)

Not technically a camera, but since the digital revolution these small Secure Digital (SD) pieces of plastic encased, non-volatile, solid-state memory have allowed us to carry thousands of pictures around with us, instantly banishing reams of 'physical' film to a distant memory. A great camera-related revolution, fully deserving to be included in our top ten. SD cards come in three distinct sizes and four families.

The original Standard-Capacity (SDSC), the High-Capacity (SDHC), the Extended-Capacity (SDXC), and the SDIO, which combines input/output functions with data storage. HC cards allow up to 32GB of onboard storage, while XC cards can offer up to a staggering 2048GB. Cards are also categorized by their data transfer and recording speed, or 'Class'. Class 2 is used for SD video recording, Class 4 and 6 used for 720 HD video recording, Class 10 for Full 1080 HD video recording and consecutive recording of HD stills, and the very fastest, classified as UHS Speed Grade 1, used for real-time broadcasts and transferring very large HD video files.

Things can only improve with this technology, so expect faster and higher capacity cards in the not-too-distant future.

SD Cards (1995-present)

THE NEW INDUSTRIAL REVOLUTION?

On *The Gadget Show* we not only review and showcase the latest technology, but whenever possible use the same tech as an integral part of making the show. During last year's 'personal helicopter' build, the Robo Challenge team used 3D printing techniques to create bespoke gears and linkages that would otherwise have to have been traditionally machined at considerable cost. Jason Bradbury and Julia Hardy also used rapid prototyping to create a mock-up of their own revolutionary modular gaming console. This method of manufacturing is at a critical juncture, but what are its origins? What is actually possible using it right now? And what will become routine in the future?

3D PRINTING 101

Turn the clock back ten years and even the idea that one day we could, at the mere press of a button or click of a computer mouse, manufacture three-dimensional items from a computer-aided design (CAD) at home would have seemed the stuff of science fiction.

The technology existed for sure, but it was both expensive and technically demanding and was firmly located in the hands of scientific research institutions along with those at the very forefront of modern manufacturing. The term 3D printing (or to be technically correct, 'additive manufacturing') can be best described as 'producing a three-dimensional object of virtually any shape or form from a digital model'. The additive aspect refers to the small layers of material that are laid down in different shapes by the printer during the printing process. This differs

Robo Challenge putting 3D printing to good use on a recent Gadget Show build

greatly from many traditional manufacturing techniques that typically remove material from an object in order to create its final shape. The joy of modern 3D printing is that it allows designers and developers to go from a flat screen image to an exactly produced part in a matter of hours, if not minutes.

A variety of processes exist for the depositing and binding of the materials used in 3D printing, but all involve the melting or fusing together of a filament or powder.

SAVE THE PLANET, CREATE A PROCESS

The spur to develop the technology that underpins 3D printing is to be found in industrial and environmental legislation passed in the 1980s. At the time many high volume production techniques used vast quantities of solvents, something governments – both local and national

– wanted to reduce significantly. In 1984, the first 'Rapid Prototyping' system was developed in the United States at the University of Texas, and what we now consider to be the recent 3D printing revolution was kick-started.

The scale of application of current 3D printing technology and its use in both rapid product prototyping and final production is almost without limit. From architectural design, automotive, civil and Industrial engineering, to the aerospace industry, the military, biotechnology, fashion, and even food, it is making major inroads. Within the next decade 3D printing tech could become as commonplace in the modern home as boring household appliance such as the microwave and steam iron.

At the moment it is not a cheap method to produce large quantities of objects at home, it is however the perfect way to produce one-offs, pre-production prototypes or even limited production runs of an exclusive item like pieces of

Rachel and Grant with *The Gadget Show* personal helicopter build that utilised bespoke 3D printed gears

jewellery. Away from the home, some of the most important uses of 3D printing are in the medical and dental industries. Surgeons can take scanned data, produce highly accurate 3D physical replicas of the parts of the patient's body that need to be operated upon, and, in so doing, see first-hand structures, shapes and form before a scalpel is wielded in anger.

In 2012 a commercial 3D printer was used in France to create a full replacement of a lower jawbone in titanium for a successful and ground-breaking medical procedure. It is quite staggering what is possible now; and what could well become possible in the future.

HOW IT IS DONE. OR THE ART OF THE CROSS-SECTION

The 3D printing process is based on a machine taking a series of digital cross-sections of an object from its CAD drawing or electronic blueprint. These are the guidelines that the software and hardware combine together to enable printing, and the most common software that sits between the original CAD software and the machine used for 3D printing is called stereolithography (STL) files.

These files take the design and create a series of triangular facets locked together to form the final shape. And just like the dots per inch employed in traditional computer printing, the smaller the facets (and, therefore, the greater the number of them) the better the quality or 'resolution' of the final object. The 3D printing machine reads the data design from the STL files and lays down multiple layers of either a liquid (often heat-melted plastic filament) or a powder of polymer or metal that is then fused into a solid during the process (referred to as laser sintering).

These gradually build up the cross sectional layers, eventually revealing the final 3D model. In units designed for the hobbyist and home user, the filament method, usually fed from a roll, is by

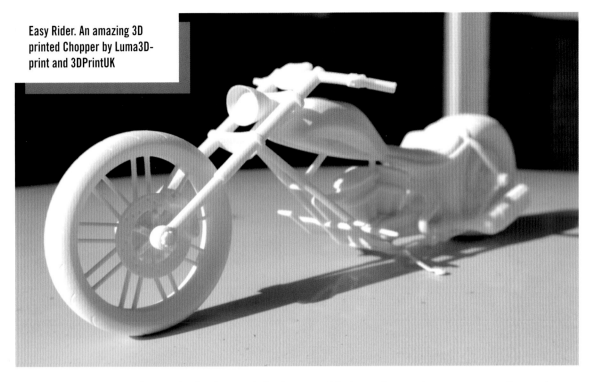

Easy Rider. An amazing 3D printed Chopper by Luma3D-print and 3DPrintUK

far the most commonly used. The layers (which correspond to the virtual cross sections from the CAD model) fuse together and eventually create the object's final shape.

It is reported that the 3D printing sector generated sales in excess of £2.3 billion globally in 2013, and this will rise further to an estimated £3.5 billion per year by 2015. The technology is only just reaching the high street consumer as the cost of the systems continues to fall, whilst initial build and setup time, output quality and range of materials have all greatly improved. You no longer need to be either an engineer or professional designer to generate the right file formats, and open source designs readily available online (for free) simplifies the process for non-CAD users.

All these factors have converged together to make the technology increasingly affordable and accessible.

The 3D printed 'click together' SULSA aircraft with its Spitfire-like elliptical wings

Almost anything is possible – a 3D printed skateboard by Sam Abbotts

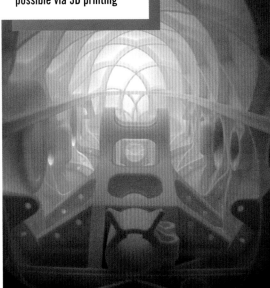

This complex one-piece internal structure is only possible via 3D printing

A BRITISH SHOWCASE OF POSSIBILITIES

In late 2013, London's Science Museum mounted an exhibition to demonstrate what 3D printers are currently capable of and what they may be able to deliver in the future. On show were printed objects including engineering components such as ultra-lightweight hinges for aircraft; works of art –especially sculptures – and practical medical applications like replacement teeth and synthetic bones. The exhibition demonstrated how diverse the scope of 3D printing is. One of the major benefits of the technology is that it allows certain geometrical shapes to be fabricated that otherwise would have been impossible to make in a single step. Laser sintering techniques, for example, can make the assembly of parts such as chains in one continuous process. Originally plastic parts produced by the method were often weak and brittle, but as the technology has advanced it has allowed more robust plastic-based materials such as nylon to be employed, which in turn offers characteristics such as heat and chemical resistance, improved wear and abrasion characteristics and very low friction.

As the technology evolves further, the range of domestic printable materials will continue to expand, offering 3D printing enthusiasts and designers the ability to cherry pick and match materials that have perfect properties to the final design needs of the printed object.

THE CLIP TOGETHER 3D PRINTED AIRCRAFT

In the fields of higher education and scientific research, many UK-based teams have recently made huge leaps forward in the use of the technology. The Computation Engineering and Design (CEDG) research team at the University of Southampton is just one of them. The group has a strategic research link with Rolls-Royce who have committed to funding long-term research into the field, including the potential use of additive manufacturing technology within full sized aircraft engines.

Researchers at Southampton recently created a technology demonstrator called SULSA, the world's first 3D printed aircraft. SULSA stands for Southampton University Laser Sintered Aircraft and the project was undertaken in collaboration with 3T, a rapid prototyping company based near Reading. Led by Professors Andy Keane and Jim Scanlan, the design team embodied some structural and aerodynamic features borrowed from a couple of famous historical aircraft.

First, SULSA uses elliptical wings (most famously seen on the WW2 Spitfire). Elliptical wings give aerodynamic advantages but are notoriously problematic to manufacture. Because of this the Spitfire was both difficult and expensive to produce (costing roughly twice that of its contemporary, the German Messerschmitt BF109). Second, the internal structure of SULSA used an arrangement of stiffeners similar to that found in a second wartime British aircraft – the Vickers Wellington. The Wellington's internal structure was renowned for its strength and ability to cope with battle damage. As with the Spitfire, this elegant design solution had two big manufacturing drawbacks, firstly the huge cost of the sheet aluminium material employed, and secondly the amount of man-hours the complex construction took.

By employing additive manufacturing techniques in the fabrication of the SULSA aircraft, both the elliptical wings and internal stiffening structure were achieved without an unreasonable time and cost penalty. With 3D printing, even at this level of sophistication, additional complexity comes largely free. There were several other design innovations in the SULSA aircraft: the aircraft did not employ any fasteners and the structural parts clipped and 'locked' into place; and the control surfaces (ailerons and tail surfaces) were printed as fully assembled multi-part mechanisms.

Just think of the possibilities that a prototype aircraft like this opens up. A computer file could be sent anywhere in the world that has the correct 3D printing hardware in place, and an aircraft like the SULSA can be printed, clipped together without tools and an avionics package added – resulting in an operational aircraft ready to take to the skies.

The one-metre wingspan SULSA aircraft is a high performance design with top speed of over one hundred miles per hour, a good payload carrying capacity and a flight time of nearly forty minutes. This is comparable to an aircraft with a much more expensive carbon fibre structure, even though the nylon material used is far less structurally efficient than a composite. What additive manufacture makes possible is the

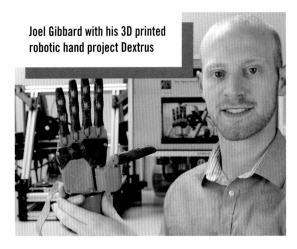

Joel Gibbard with his 3D printed robotic hand project Dextrus

addition of hundreds of delicate structural stiffener elements arranged in an efficient pattern.

Even though sophisticated CAD tools are used to design such parts, it remains a labour intensive process to add these types of features. Essentially the time required to create a part is proportional to the individual part's actual complexity. This is an important area where the Southampton team is doing significant research. The goal is to develop 'intelligent' design tools that automatically add the necessary complexity to parts to specifically suit the additive manufacturing processes. It is an exciting time and the United Kingdom is at the very forefront of the field. Technology developed in this and other research programmes will filter down to both industry and consumers, benefiting everyone.

PRINT YOUR OWN . . . ANYTHING

The application of this technology in all walks of life is only limited by the size and speed of the printing machines themselves. Currently, the process can take many hours, even days for large complex objects, but as printer technology evolves, the speed at which these machines will operate will improve. In a world where '3D Print Stations' are universal, the day may come when, instead of buying something of the shelf, you pay for the build file and print/manufacture it locally.

This will have a huge knock-on effect on regional and global economics, eliminating the cost of inventory and distribution. It will still be about getting the item, whatever it is, to exactly the right place and at the right time, but will involve a whole new mindset and restructuring of many consumer industries – equivalent, in many ways, to the impact that digitisation has had on the publishing and music sectors.

As things stand, 3D printing techniques have constraints that need to be addressed when designing objects. In the case of laser sintering, the designer must ensure that the unused raw material powder can be entirely removed from the finished part; this means that the object cannot be designed with 'closed' spaces. Even with an 'open' structure, the powder might be difficult to remove, so labyrinthine structures generally need to be avoided. Since SULSA, the Southampton team has designed a full-sized unmanned aircraft called SPOTTER (Southampton Platform for Observation Tracking Telemetry and Environmental Reconnaissance), a drone that makes use of much of the technology that was pioneered in SULSA.

The aircraft has been designed as part of a European project to demonstrate the feasibility of using unmanned autonomous aircraft for patrolling coastal waters to undertake pollution monitoring, and for the detection of illegal activities on behalf of the maritime police forces and port authorities.

The aircraft is designed to carry a five-kilogram sensor payload and has a surprising endurance of five or more hours. It is designed to be super-reliable and capable of flying in poor weather. Its central fuselage is a large laser sintered part. The additive manufacture of this structure builds in a full range of complexity, meaning that it is ready to have the main fuel tank, centre spar housing and payload pod attachment located during assembly – something that would not be possible without 3D printing technology.

This process, being pretty much tool free, permits 'zero-cost' customisation of parts, so that the design team can develop various versions of the aircraft, with different fuel tank capacities for example, depending on the use the aircraft is going to be put to. As for the future, a recent experimental programme by the Southampton team is designed to develop processes that will 'print' electrical and electronic components. Their vision is to be able to print fully functional mechatronic devices using a single process.

Recent results in the work have shown that conductive tracks and the full integration of electronic devices in the context of a laser sintered structure is possible. This would be ground-breaking, as, in theory, one could print not only the aircraft's structure, but along with it the appropriate avionics package. Now, if the two technologies are then somehow combined into one additive printing process, then the era of the clip together, 'plug and play' drones, sent as one file to a 3D print station anywhere in the world becomes viable.

TOUCH WHAT YOU SEARCH FOR

Yahoo! Japan recently unveiled 'Hands On', a voice-activated 3D printer in a cloud-like housing that's permanently attached to their Internet search engine. It has been built with a voice-activated user interface and initially resided at the Special Needs Education School for the Visually Impaired in Tokyo. Students were able to press large buttons and verbally instruct the machine to (in theory) make whatever the students wanted, right there and then. After a few minutes, the machine would dispense the objects in the form of palm-sized plastic miniatures.

This functionality, making the dreams of these children become a physical, albeit at a reduced-scale, reality, Yahoo is calling the 'future for the Internet.' Yahoo! Japan poses an interesting question on its website: 'The Internet is visual and auditory. What if the sense of touch became possible? What does the future look like?' That's a really tough question to answer, but additive manufacturing is looking like a technology that could play a part in suppling part of the answer. Interestingly, if a student's search yields no results, Yahoo! Japan then puts an advert on the project's

rks of art like this intricate
ll design showcase what is
rently possible

main landing page in an effort to obtain the open source file from the current 3D printing community.

The machine will stay at the school for the timebeing after which Yahoo! Japan will send it to various institutions around Japan. Yahoo! Japan also said it plans to eventually make the technology available to the general public, opening up the possibility of voice activated 3D printing in our own homes sooner than we might expect. OK, so it is not as sophisticated as the 'Replicators' on Star Trek, but if we can 3D print objects like this now, who is to say Star Trek-like food fabrication isn't a possibility? Imagine 3D printing a cake, layer by layer, then following up with printed icing: a decade ago the concept would have been ridiculed, but now it is at least conceivable given the tech at our disposal.

3D printing is an increasingly important tool for the research community, manufacturers needing accurate, fast but cheap (well at least not ruinously expensive) prototyping and tech-savvy consumers. What really brought the potential of this technology to the world's attention was the controversy surrounding the demonstration in the US that any individual could print a fully working gun from easily sourced online blueprints. Search YouTube and you will quickly see the wide range of possibilities this component of a potentially new industrial revolution is capable of and the infinite possibilities and applications it will be capable of in the not too distant future.

3D printers have been used to create a fully functional robotic arm, a 3-Axis camera mount for a RC quad copter, click together phone cases, and, remarkably, an entire dress for a fashion week. NASA has widely talked about using the technology to make components on demand at the orbiting International Space Station. It makes complete sense and would be far cheaper than the ruinous payload launch expense of objects manufactured on *terra firma*.

TRULY LIFE CHANGING TECHNOLOGY

In its simplest form 3D printing offers the chance to make a replacement battery cover for a television remote or a complete case for a mobile phone. On another level, visionaries like UK-based roboticist Joel Gibbard are employing the technology in ways that will literally change lives. Joel is currently in the process of designing and developing a 3D printed robotic prosthetic hand he has called 'Dextrus'. The Open Hand Project, as with many recent tech initiatives, is being crowd funded, and the 3D printed, open source prosthetic/robotic hand will be licensed for free, patent-free. The Open Hand Project is raising money on indieGogo for the first production run and a pledge of £460 will get you a fully assembled hand; for £700, you will get the hand complete with all the electronics. This amazing project's main goal is to make advanced technological prosthetic hands more accessible to amputees, but that's just the tip of the 3D printer iceberg.

Joel is actively encouraging others with access to 3D printing technology to do some good within the global tech community too. Being open-source, it means all of the blueprints to actually make a robotic hand will be published online without patents. Anyone can manufacture his or her own example. This simple and ethical offering will ensure that the Dextrus hand can be manufactured easily in developing countries, keeping the costs low and affordable in parts of the world where importing such a life-changing traditionally manufactured object might be prohibitively expensive unless given significant governmental subsidies.

But that's not all; the project goes beyond amputees. It is ideal for integrated use in commercial projects such as bomb disposal robotics or adding gestures and even sign language that help relay information better in telepresence robots. At the relatively low level of cost, the technology could even be put to good use by the DIY roboticist who wants to create a humanoid robot on a budget.

Joel says: 'The Dextrus robotic hand offers much of the functionality of a human hand. It uses electric motors instead of muscles and steel cables instead of tendons. 3D printed plastic parts work like bones and a rubber coating acts as the

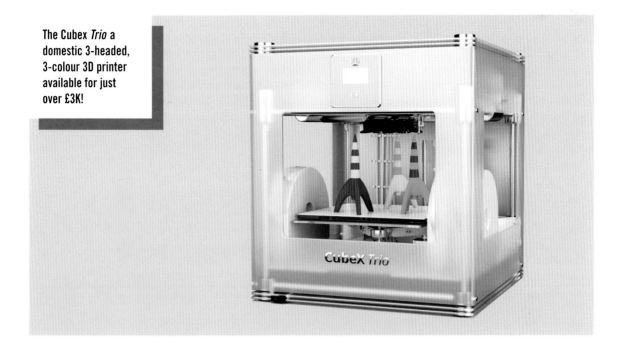

The Cubex *Trio* a domestic 3-headed, 3-colour 3D printer available for just over £3K!

skin. All of these parts are controlled by electronics to give it a natural movement that can handle all sorts of different objects.'

3D printing technology is also helping surgeons perform delicate reconstructive surgery. Scientists at Queensland University of Technology have recently printed what can only be described as bespoke biodegradable scaffolds that can patch damaged skulls and stimulate new bone growth. The same team has also revealed that they are developing similar scaffolds to grow body parts such as ears and noses. In the UK, John Hunt from the University of Liverpool recently made a 3D printed bladder using actual cells. The possibilities seem endless, and in time reconstructive surgery performed by a tissue forming 3D 'bio printer' (as seen in Luc Besson's film *The Fifth Element*) will become a reality.

THE REVOLUTION HITS THE HIGH STREET

Not since the Spinning Jenny became the catalyst for the Industrial Revolution has a piece of new technology fostered such debate (and hype some might say). The early adopters of 3D printing claim that it has the ability to change the world, some say it already has done so. Two of the UK's best-known high street electronics and technology retailers have already begun selling consumer-friendly 3D printers – the 'Up! Plus' at £1,600 and the 'Cube' at £1,195. And economically speaking, it doesn't quite end there as the 3D printer cartridges cost a further £50 or more for roughly 700g of plastic. Not exactly cheap, but it is an amazing price considering the technology that is involved; six months ago most 3D printing machines that worked at this resolution would have cost four times as much.

The Cube has been described as one of the easiest to use models on the market; you can print straight from the box as it is supplied with a selection of 3D print files. It easily interfaces with user generated STL files or any of the freely distributed plans from the ever-expanding number of 3D print forums and enthusiasts' websites. It uses material cartridges of recyclable and compostable plastics that are available in sixteen different colours including vibrant and neutral colours, metallic silver and even 'glow in the dark'.

If you want to print larger items and in multiple colours, then 3D print specialist online retailers can also provide units like the £3,195 CubeX *Trio* desktop 3D printer. It currently has the largest print volume in its class for a domestic 3D unit with the ability to print objects as large as a basketball! 3D printer tech is a good example of technology that is at the cutting edge one minute, and part of the gadget-mainstream the next.

Industry experts have compared our current enthusiasm for 3D printing with the craze for bread-making machines in the nineties. Thousands of them were purchased as part of a domestic gadget trend, but just months later most ended up in the back of cupboards or were banished forever to the loft. Whether we use 3D printer technology to create something life changing, or just a piece of unusually shaped plastic, the decision is ours and that's perhaps the point. We can now manufacture our own prototypes and even final components, sometimes in a matter of minutes, more often in a matter of hours. But we control the process; we initiate it and are the recipient of the final product. The might of being able to manufacture is now in the hands of the consumer; we are no longer at the mercy of the factories and the 'vision' of product marketeers. Perhaps what was the true driving force behind the first industrial revolution – the industrialization of the manufacturing process – is now being threatened by the latest one.

Small batches of items like these smart phone cases can be printed in your own home

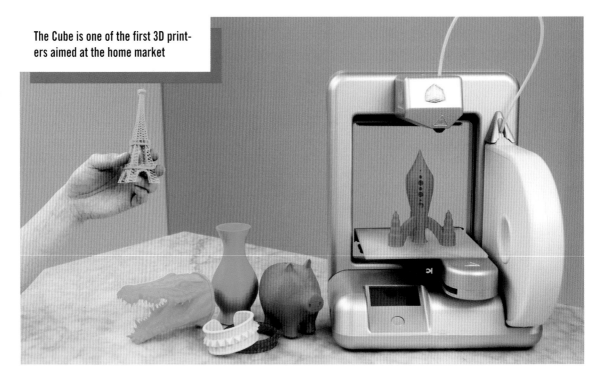

The Cube is one of the first 3D printers aimed at the home market

ASTOUNDING GADGETS

Our tech-dependent world is the direct result of hundreds of years of scientific enquiry. Some of the greatest minds in history have looked at the natural world and asked the question, 'how does it all work?' Theories – some good, some bad, some plain lunatic – have come and gone as science has peeled back the onion layers of understanding. The application of science means that today we benefit from unimaginably sophisticated technologies: HD televisions, tablets and smart phones, advanced medical diagnostic tools such as MRI (magnetic resonance imaging) scanners, electric cars and the Internet. All of this thanks to a succession of brilliant questioning minds and the uses to which their insights have been put.

The Renaissance was the time during which science became increasingly based on observation and measurement. It gave birth to the Age of Reason (or the Enlightenment) in the late seventeenth and early eighteenth century, when a new age of individualism and reason, rather than collective reliance on tradition and dogma, produced an era of intellectual thought that sought answers by reducing the study of the natural world to a collection of laws. The new scientific rigour was helped by an increased willingness to question religious orthodoxy, and its luminaries would include Robert Hooke, the leading experimenter of his age and Sir Isaac Newton whose classical physical laws still find daily use today.

By the beginning of the twentieth century, physics had entered a period of post-Newtonian revolution.

The discovery of the nature of the atom by Ernest Rutherford in 1911, when he showed that it consisted of a positively charged nucleus surrounded by negatively charged electrons; Einstein's monumental and paradigm-shifting General and Special Theories of Relativity that included the famous equation $E = mc2$ revealing mass and energy to be one and the same thing; the emergence of quantum mechanics that told us electromagnetic waves behaved like particles and that matter can be seen behaving as a wave (known as wave-particle duality) and Heisenberg's 'uncertainty principle' of 1927 (the impossibility of measuring a particle's position and momentum simultaneously) that meant all of our assumptions about one thing 'causing' another had to be discarded and would have Einstein protesting in despair, 'God does not play dice with the universe'. Nevertheless, quantum mechanics, for all of its counter-intuitive weirdness, has become an indispensable tool in the investigation of the realm of the sub-atomic.

All of this three-hundred year intellectual endeavour has been driven by our species' profound, unquenchable thirst for knowledge and understanding. If there is any point to our existence, then surely it is to explore the fundamental nature of our universe and to try to answer the deepest questions about space, time and matter. How did it all begin? What is it all made of? How might it end one day? Could it be any other way and why is it here at all?

Great cathedrals of knowledge-seeking are being built and operated – from the subterranean world of the Large Hadron Collider (LHC) where the sub-microscopic meets humankind's most gargantuan engineering enterprise, to the fabulous space-located telescopes that are busy discovering thousands of exoplanets outside our solar system, some of which may harbour life. And confirmation of that discovery alone – that life exists elsewhere in the universe – would surely justify every single penny spent in the pursuit of answers to all of the scientific questions posed since Archimedes jumped into his new bath. And of course this isn't just curiosity for curiosity's sake; all of our modern life-enhancing, life-supporting gadgets and tech come from the application of pure scientific research. It is estimated that at least twenty percent of the world's business is done over the web and CERN, the home of the LHC, invented the web, which in itself is enough to justify the expense of so-called esoteric research for several hundred years.

THE LARGE HADRON COLLIDER

It set out to answer one of the biggest questions in science. At the moment of the Big Bang, 13.7 billion years ago, the universe came into being and very quickly some of the immense energy released condensed into fundamental particles. The process of creating 'stuff', the matter we see all around us, had begun, but that process required some of these particles to acquire 'mass'. Without it the universe would still consist of nothing more than a cloud of fast-moving particles, and none of the clumping together of these particles to form neutrons, protons, and atoms would have occurred. Lacking the property of mass there would be nothing for gravity to 'work on', so no stars would exist, nor for that matter would any of us. If an initial hot plasma of newly forged fundamental particles arose out of a vacuum, how did a percentage of these particles actually acquire their all-important mass? What is mass and how does it make stuff, stuff?

In the mid-1960s a British physicist, Peter Higgs, proposed an elegant theory that offered a solution to the problem. Today, the best explanation we have of what matter is and how it behaves is known as the Standard Model of particle physics; it is a powerful description of a small menagerie of fundamental particles and the forces that act upon them. The particles can be said to broadly occupy two distinct camps: fermions, which include quarks (they come in various guises and stick together to form the neutrons and protons that in turn make up atoms and molecules) along with electrons; and bosons that are force carriers – they mediate various types of interactions between the particles, and give rise to what are known as the strong, weak, and electromagnetic forces. Here's an example: perhaps the best-known boson is the photon (the quantum of light and all other forms of electromagnetic radiation) that mediates the electromagnetic force between electrically charged particles. This mediation requires the presence of a field that permeates the whole universe in every direction, and it is the perturbation or disturbance of the field that gives rise to the forces 'felt' by the fermions. Photons (they bring us the light from stars and give structure to the atoms and molecules that we're all made of) have no mass, and it is this massless property that explains why the electromagnetic force is felt over distance, whereas the weak force that drives the energy creating processes in stars only operates over very short distances. The weak force is carried by the W and Z bosons and they do have mass, quite a lot of it. The point being that if mass is a fundamental property of the universe, this explains why the force mediated by photons has an infinite range

The Big Bang

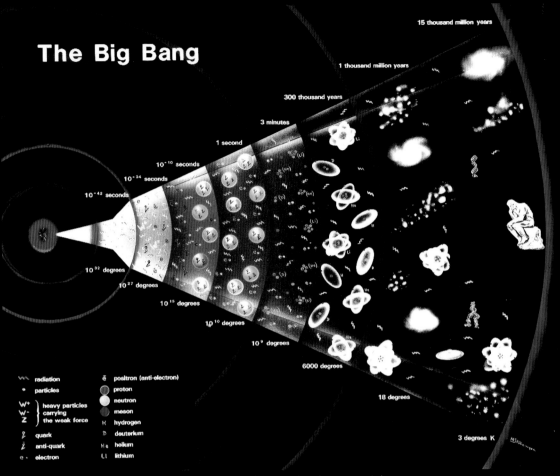

© and courtesy of CERN

A schematic showing how the Big Bang eventually cooled down and became the universe we see today.

while that mediated by W and Z bosons is short range. It is similar to what woud happen if you compared throwing a cricket ball with hefting a 5kg medicine ball. The lighter (lower mass) cricket ball is going to go a lot further! What Peter Higgs proposed was that there was a mediating boson and a field that would confer mass on some fermions. As the fermions moved through the field they would feel a kind of tugging resistance imparted by the field and the disturbed field would 'emit' a boson. These were named the Higgs field and the Higgs boson.

To prove the theory would require an experimental facility to be built that would be unlike any that had been attempted before. The problem was that the predicted Higgs boson was understood to be an amazingly short-lived creature, existing only fleetingly in the real universe and whose existence would only be detectable in the debris of incredibly high-energy collisions of fundamental particles. The conditions during these engineered collisions would have to replicate those only seen a fraction of a nano-second after the Big Bang.

Particle physicists have been using machines to smash things together and poke about in the debris for a generation, but the energy requirements and the subsequent colossal volumes of data needing to be analysed in order to search for evidence of the Higgs boson would be orders of magnitude greater than anything attempted previously. It would involve designing and building the biggest and most sophisticated machine humankind has ever conceived. And it would be nearly fifty years since Higgs first proposed the existence of the

history of the universe

PRESENT

15 BILLION YEARS

5 BILLION YEARS

First Supernovae

Galaxy and Star Formation

1 BILLION YEARS

THE UNIVERSE BECOMES TRANSPARENT

Decoupling of Matter-Radiation | Formation of Atoms | 10^{13} sec.

PHOTON EPOCH

Nucleosynthesis of Helium | 10^{2} sec.

Disappearance of Positrons

LEPTON EPOCH

Confinement of Quarks Formation of Protons, Neutrons
Disappearance of Antiquarks | 10^{-10} sec.

ELECTROWEAK EPOCH | QUARK EPOCH

Asymmetry Q - Q̄ , L - L̄ | Magnetic Monopoles? | 10^{-34} sec.

Cosmic Inflation?

GRAND UNIFICATION EPOCH

10^{-43} sec.

QUANTUM GRAVITY EPOCH

Big Bang

MATTER DOMINATED ERA

RADIATION DOMINATED ERA

Spiral Galaxy

WE ARE HERE

Heavy Atom

Black Hole

Heavy Star

Protogalaxy

Hydrogen Atom

Helium Atom

Helium Nucleus

The Desert

microcosm CERN

© and courtesy of CERN

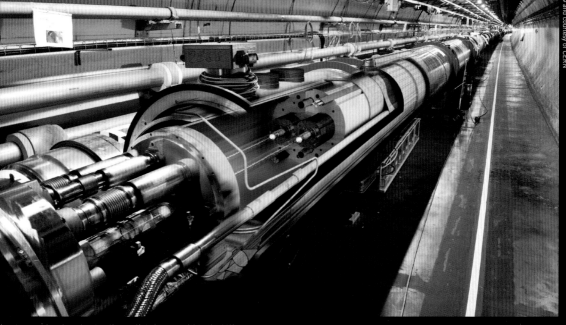

© and courtesy of CERN

A cut-away drawing showing the particle beams within the LHC.

new particle before the Large Hadron Collider, located at the CERN nuclear research centre that straddles the Franco-Swiss border, would be ready to start searching for the elusive Higgs boson in earnest.

It took thousands of engineers and scientists twenty years to design, and a further ten years (plus £2.6 billion) to build, the underground 17-mile tunnel and the particle accelerator into which beams of protons would be fired and accelerated to 99.9% of the speed of light. Two beams circulate the ring in opposite directions before being periodically smashed violently together at a combined speed of 600 million mph. When operating, the tunnel is maintained at temperatures colder than outer space (-271.25 °C) by huge reservoirs of liquid helium; 9,000 immense niobium-titanium superconducting quadrupole electromagnets bend the billions of positively charged protons around the tunnel, which they circulate 11,000 times a second. Dotted around the 17-mile machine are colossal detectors connected to a supercomputing grid ready to sift and analyse the mountains of fiendishly complicated data that is generated when the beams collide.

In July 2012, two experimental teams at CERN operating different detectors, one the 7,000 ton ATLAS (A Toroidal LHC Apparatus) and the other the 12,500 ton CMS (Compact Muon Solenoid), announced that they had found evidence of a new particle consistent with the predicted characteristics of the Higgs boson. The 45-year hunt was over – possibly. More accurately, the first step in locating and measuring the properties of the elusive particle had finally borne fruit. The CMS team claimed that they had seen a 'bump' in their data corresponding to a particle 'weighing' approximately 133 times more than a proton. Particle physicists measure the confidence they have in their data in statistical terms and, while this was only an early indication that a particle had been found, it was located at the mass-energy range expected for the Higgs boson with a one-in-two million chance of error. The Atlas results were even more promising, claiming data for a particle at a slightly higher mass-energy (heavier) than the CMS observation. The teams had been working independently and were blind to each other's progress.

In March 2013, the by now collaborating ATLAS and CMS scientists announced further results in Italy that confirmed the 'magnificent' discovery. The Higgs boson had been snared. However, the Higgs boson has only just begun to give up its secrets. The question that

© and courtesy of CERN

On the surface CERN looks like any campus-style academic institute. It belies the cathedral of science that sits hundreds of metres below ground.

needs to be answered now is whether this particular particle is the actual Higgs boson first postulated in the mid-1960s or some other lighter boson predicted by alternate theories. The widespread belief is that the observed particle is the long-anticipated Higgs boson, and this was good enough to see the Nobel Prize awarded to Peter Higgs in 2013.

The Large Hadron is currently in the middle of a planned shutdown for maintenance and upgrades. When it comes back online in 2015, it will be operating at nearly twice the energy levels (14TeV), allowing the experimental teams to smash particles even harder and gain considerably in 'resolution'. The Higgs boson will be scrutinised mercilessly.

The question of how particles acquire mass may have been answered, but physics has many equally profound mysteries to resolve. Part of the universe seems to be missing. Most people will know by now that the matter we can see making up the dust, planets, stars and galaxies of our universe (so called baryonic matter because it is made fundamentally from quarks) only makes up four percent or so of the required mass-energy of the universe. Dark matter (we can't see it or measure it as it resolutely refuses to interact with 'normal'

matter) may be made from the heavier siblings of fundamental particles we already know about and the LHC may give us our first glimpses of this mysterious dark universe. And then there is the peculiar and very poorly understood matter-antimatter imbalance we see in our present-day universe (they were created in equal quantities during the Big Bang so we shouldn't be here at at all as each particle should have been annihilated by its anti-matter equivalent), something that will be studied by the LHCb (Large Hadron Collider beauty) detector.

Albert Einstein said: 'It stands to the everlasting credit of science that by acting on the human mind it has overcome man's insecurity before himself and before nature.' Since we first looked up at the night sky and wondered at the majestic sweep of the Milky Way, humankind has sought answers to the most profound questions posed by the universe, always seeking an understanding of our place in it. The Large Hadron Collider stands at the very summit of human ingenuity; it is the culmination of our evolved big brains, and is helping to decipher the most fundamental mysteries of our universe and the forces that govern it. It is an awe-inspiring achievement and we should continue to celebrate it and the work that it does.

THE TIANHE-2

The world of modern gadgetry has the computer at its heart. Developments in processing digits have given us everything from tablets and smart phones to digital cameras and Ultra HD TVs. So the world's most powerful supercomputer could well claim to be the world's ultimate gadget.

It's a very hotly contested area. The Top500 project produces a ranking of the world's most powerful computers every six months and, China's Tianhe-2 led the June 2013 list. It was built by the National University of Defence Technology and is operated by Guangzhou's city government. The city paid for most of its cost, about 2.4 billion Yuan or £244 million.

The statistics are phenomenal. Your laptop's processor may be quad core but this supercomputer has no less than 3,120,000 cores and 1.4 petabytes of RAM. Peta, incidentally, means ten to the power of 15. It can achieve more than 100,000 times 300 billion calculations per second. Or, put another way, 54.9 petaflops. An entry-level laptop manages around 0.9 gigaflops. Tianhe-2 translated into English is Milky Way 2 – a galaxy with 300 billion stars and other impressively high vital statistics of its own.

All this computing power requires considerable electric power. Running flat out, the colossus consumes 17.6 megawatts of electricity, equivalent to the electricity demands of about 6000 homes. And top computers like the Tianhe-2 require a well-ventilated basketball court-sized room in which to perform their super calculations.

The Top500 list defies those who predicted the imminent demise of Moore's Law, the dictum that the power of computers doubles roughly every two years. The power of the computer that holds number one position in the rankings has been rising in line with Moore's Law since the project started back in 1993.

However, even these super-powerful, supercomputers need to be put into perspective. For a start it is by no means certain what Tianhe-2 can actually be used for. There are tasks such as modelling climate and nuclear reactions, and research into quantum mechanics and spacecraft aerodynamics, which would suit all this computing muscle. In practice it's much easier to build the hardware than write the software to exploit it. It would take years of code writing by expert teams to make using such a powerful machine worthwhile.

The Tianhe-2 is currently the world's most powerful supercomputer.

The Tianhe-2 consumes enough electricity when running at peak operating levels to power 6,000 homes!

Seemingly complex tasks such as simulating car crash tests, developing new drugs or creating movie special effects are actually achieved very well with less powerful machines. It is human programming skills, not computing power, that form the barrier. Others talk of Tianhe-2 revolutionising China's rapidly growing car industry but here the needs are more the traditional ones of better design, quality and creativity than computing power.

Tianhe-2's power, though huge, is nowhere near matching the capacity of the human brain. And the components aren't revolutionary. Those computer cores are standard Intel Ivy Bridge and Xeon Phi chips. It's the same basic technology as you'll find in your everyday computer.

Maybe the next computing breakthrough will be a device that isn't a computer as we know it at all, perhaps some sort of biological computer running on biomolecules like enzymes and DNA. Possibly even a quantum computer or one that uses nano-sized magnets. Aside from whether any of these developments might soon challenge the world's current top supercomputers, one thing is certain – the Tianhe-2 won't be the world's fastest computer for long.

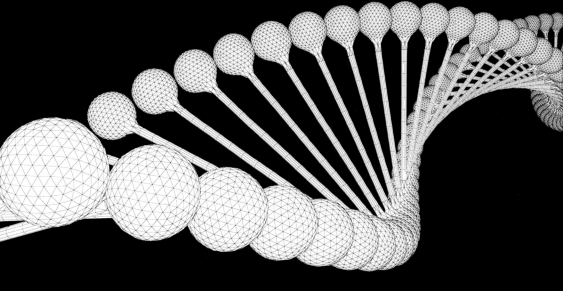

DNA MEMORY STICKS

One of the biggest problems in modern life is keeping your digital data alive. Years ago we all shoved our family photos in the kitchen drawer. When we wanted to relive a few memories, even decades later, we could pull them out and they'd still be there, faithfully waiting for us. These days life is much more complicated.

Digital data is, by its very nature, fragile. Hard drives fail, memory cards get corrupted or lose their charge, and optical discs succumb to the dreaded bit

rot. Some of my recorded DVDs have become unplayable after only a few years. One answer is to make lots of copies, and make new files every few years as the copies age, to keep them alive. But life's too short to spend it forever backing up files.

And even if you do manage this extraordinary feat there's the problem that formats change with incredible rapidity. Can you still put your hands on a floppy disk drive? Can you still gain access to that treasured recording on an audio cassette or VHS when the player's broken?

For many, cloud services are a solution but they're expensive, and not totally reliable or secure. The problem's getting worse as our demands for data storage are growing almost exponentially; every extra hike in the definition of our stills and video, for example, means bigger files and more to store. Ultra HD 4K home video recording is now a reality. That means still bigger files and more storage problems.

What if there could be a data storage solution that was safe, that was cheap and one that could cope with our ever-growing demand for data? Well, there could be, and it uses DNA. The material that stores details of life and how to reproduce it could also be a solution to our data storage needs. Our data is, quite literally, transferred from the hard drive to the test tube.

It has lots of advantages. It's very compact. Researchers from Harvard University reckon that every bit of data humankind produces in a year could be stored in just four grams of DNA. Nick Goldman and Ewan Birney of the European Bioinformatics Institute (EBI) near Cambridge think all of the world's digital information ever created could fit into the back of a lorry. The Cambridge researchers have chosen ternary rather than binary storage - storing information as 0s, 1s and 2s rather than just 0s and 1s - which has helped to solve an accuracy problem with previous DNA storage methods.

As long as it's kept in reasonably dark and dry conditions the data is incredibly durable. DNA information has been retrieved from creatures that died thousands of years ago, like Neanderthals and Woolly Mammoths, so it should stand up better to time than anything we've thought of to date.

DNA storage has its limitations though. It's currently very expensive as well. The last Cambridge experiment encoded just 739 kilobytes of data including an MP3 of Martin Luther King's 'I have a dream' speech and a PDF of the paper by Francis Crick and James Watson describing the structure of DNA, and researchers reckon the current cost is about £850 a megabyte. But the cost is coming down at a very fast rate and it's hoped that it will be competitive with existing long term data storage techniques within a decade.

Further issues are that it's not rewritable, you can't access the data randomly like you can on a hard drive or CD, and reading the data is also very slow. But the speed is improving, and as we're talking long-term data storage, speed of access and reading isn't really that crucial.

Maybe one day data could be stored in living DNA too. Then we could literally access all our files directly, after they've been logged directly into our brains.

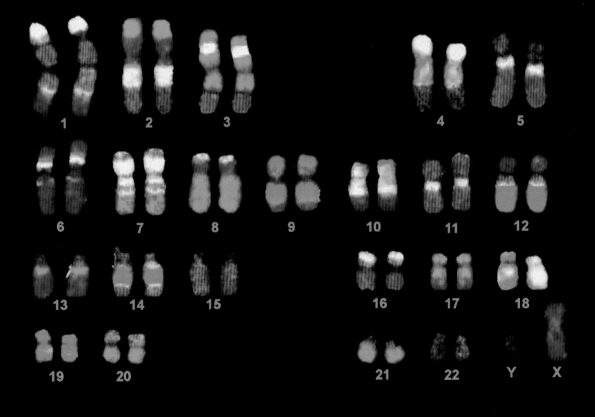

WHAT DO YOU KNOW ABOUT?
Jason Bradbury

Jason has been on *The Gadget Show* since day one and has worked as a television presenter and producer on hundreds of shows in a career spanning almost twenty years. A veritable force of nature when it comes to all things tech, his *Gadget Show* tests, often set in the most dangerous or ludicrous situations the show can dream up, are legendary. When not working on the show, Jason spends his time developing *SupaRobotAttack,* the dedicated YouTube robot fighting channel.

Here are five things you might not know about Jason . . .

1 I hold flying, driving, motorbike and amateur radio (including Morse code – call sign 2EOJAB) licenses. *So if you ever need to talk to anyone on the other side of the globe, send an SOS or be flown, driven or ridden anywhere, I am the man to call!*

2 I collect vintage synthesizers (Moog Prodigy, Roland Jupiter 4 and ARP 2600) including every individual model that was used on the 1981 Dépêche Mode album *Speak and Spell.*

3 I have no upper front teeth (at least of my own!) since an unfortunate BMX accident when I was fifteen. It doesn't help that I had another one knocked out by the pneumatic Robo Challenge Kung Fu Robot we built on the show back in 2010. Ouch!

4 My father was one of the scientific team that consulted on the plastics and their formulations used in the *Eurofighter.* How cool is that!

5 I was once a comedy partner of David Walliams. We were part of a group called *David Icke and the Orphans of Jesus.*

Google Glass

Gadget nirvana or a niche-dwelling vanity?

Google Glass is wearable tech that Google describe as an augmented reality device. It encompasses a live view of the world augmented by computer generated sensory input. It still needs to be connected to Wi-Fi or paired with a phone via Bluetooth, but with its camera (stills and video), voice recognition capability and audio, it's the start of something very big. Isn't it?

One spring day in New York, Martin Cooper, a senior researcher at Motorola, did something extraordinary for the time, and truly significant in tech history. It would take another twenty years for the consequences of Cooper's actions that day to fully resonate and to fundamentally change the way we go about our daily lives. In fact today it would be unimaginable for a large percentage of the world's population to effectively communicate without the results of his pioneering work. On 3 April 1973, using a somewhat heavy portable handset, Cooper made the first call on a handheld cellular mobile phone (to a rival at Bell Labs as it happens) and in that moment not only was a new industry conceived if not actually born, but there followed a profound recalibration of the way humans interact with each other. Ironically, Motorola, who would continue to be an innovative mobile communications company at the forefront of the communication industry for many years, would, in due course, be thoroughly upended by the arrival of the smart phone and later, in something of an ironic dénouement, sell their handset business to none other than Google.

Jump forward thirty-four years to 29 June 2007 and the launch of the first iPhone. The product of Steve Jobs' vision, a one thousand-strong research team and a one hundred and fifty million dollar spend, Project Purple (as the secret programme was called) would result in another paradigm shift in human affairs. It would take time, but not much, for this, the third element in the modified technological landscape (the second, of course, being the Internet), to impact on most aspects of the way society functions. Humans have an insatiable urge to communicate; it seems to be hard-wired into us. We are profoundly social animals and also magnificent data processors, consuming over one hundred thousand words and in excess of thirty-four gigabytes per person per twelve-hour day according to research conducted by the University of California, San Diego in 2008. Interestingly, as this research makes clear, this prodigious data consumption largely ignores computer delivered information. The advent of mobile data – from phones, tablets and phablets – will undoubtedly have significantly added to our already gargantuan appetite for information and connectivity. In summary, we are, as a species, insatiable information Hoovers, who love to share our thoughts, feelings and emotions (what else is a social network website other than an extension of our urge to gather around the nighttime hearth) and we are becoming adept, dexterous even, at embracing new tech to satisfy these deeply atavistic needs.

And now another potentially culture-shifting tech innovation is about to be set among us. 'Get technology out of the way so it can do the work for you' is a Google recurring theme and leaving aside the real or imagined privacy issues, Google Glass is either the nirvana of gadgets or the most controversial device ever; perhaps destined to be nothing more than a niche-dwelling vanity that goes the way of the Apple Newton and the Segway. But there are reasons to think not.

Is Google Glass a solution, a problem or an unfinished idea? Arguably it's all three and then some, but let's take each question in turn. A solution to what exactly? Assuming that we humans continue down the path of evolving sociability through tech interfaces – and it is a generational thing – then dissolving the up-until-now button-pushing, screen-swiping, artifact-clutching *modus operandi* is inevitable – especially given the commonality of design intention among the big tech companies and the sheer pace of development. Remember that Google X (the secretive Area 51-like research facility from which Glass has emerged) and its Samsung, Apple and Microsoft equivalents are working on concepts with five- and ten-year product development horizons. Samsung's recently announced curved screen phones didn't just appear overnight, but would have been in development for many years with some of the sharpest brains imaginable spending their every working hour on it.

So Google Glass is then, if not the first, an important product-launch step in transitioning to a time when computers are genuinely a data delivering, but unencumbered extension of the mind. We can speak to Glass, query it, command it and utilize it in a way that smart phones and smart watches only permit tangentially. A first experience of using Glass tells you that, while it is still very much a work in progress, we are entering a whole new world. There is a deep-felt excitement when you first put them on, say, 'OK Glass' and the prism screen fires up. The screen image sits in the top corner of your right eye's field of view, is non intrusive and entirely legible. The bone-conduction audio works incredibly well and here's the thing, despite its current relatively low score on usefulness, you quickly realize that something entirely natural (foreshadowed by so many sci-fi films and TV series) is actually happening: with eye movement and voice command alone, you are (pretty much) connected to the virtual world of SMS, email, Google search, and Facebook. While it may be a strange thing to say when your face is adorned by still odd-looking eyewear, it really is a case of the computer getting out of the way. Overlaying data into your personal field of view is an immensely useful thing and the menu of functions already available to users of Glass is an intriguing place to start. Although minus GPS at present (you have to pair it with your phone), directions become more intuitive and you can view translations or transcriptions of things being said more or less in real time (how Star Trek is that?!). Primarily, Glass in a notification-based system that allows you to scroll through and answer messages, all on the move. While Glass is unlikely to be the natural delivery device of rich content (Twitter is the type of app that is ideal for Glass), the fact is that anything that can be put on a screen can, in theory, be displayed in both sound and vision through Glass.

The move from pure concept to street-worn reality is posing awkward questions about social mores – mainly concerned with the acceptability of having a camera permanently stapled to your head. It is unlikely you would receive much of a welcome if you walked into a busy bar or a leisure club changing room with a GoPro camera pinned to your chest, so it is all going to take a bit of working out. Those who have been living with Glass for a year now since the first test versions (Google Glass Explorer) were distributed, make the point that Glass at least mitigates the 'apparent rudeness' of socially disengaging with peers when endlessly checking smart phones and tablets for updates and messages. Or, worse still, the usually unintended relayed signal of 'I am bored with you now' when a smart watch wearer raises their wrist to check for emails! Glass demonstrators have a neat trick when meeting a Glass novice; while making a big thing about taking a picture with Glass (you have to point you head quite deliberately at the subject and then issue a clear voice command or tap the side of the Glass – none of which tends to go unnoticed), a few moments later they show you the other picture, the one they took sneakily on their smart phone. The one you didn't notice them taking!

The projection behind the sofa shows you what you see in the top right corner of your field of view when you first activate Glass.

We swim in a data-rich sea and isolating what is important, vital even, from the chaff of superfluous and misdirected 'information' and constant digital chatter is a challenge. Handheld devices presently hold sway and have become our closest companions. However, they are incursive and can't help but create an unnatural barrier between us all in the course of our daily lives. Used with the right intentions and discipline, Glass has the potential to rewrite the way we interact – both with the incredible data flows we're subject to and, at the same time, liberate us from the constant anchor points of smart phones, tablets and phablets. Perversely it has the potential to make us better, more socially aware and friendlier citizens.

The current crop of first generation 'Explorer' Glass has limited functionality, launching with a small pool of official core-apps, but things are moving fast. Third-party Glass apps, also known as Glassware, have been around pretty much from the beginning, but lacked an adequate new app discovery ecosystem equivalent to Google Play. That is being addressed through MyGlass (the control centre and interface for Glass) and the release of the Glass developers' kit and Google's own Glassware review process. We will see a burgeoning of Glassware over the coming months. With voice command alone you will be able to upload photos to Twitter and Facebook as well as Google+, update your Facebook status on the fly and call up the history and points of architectural interest along

Certainly privacy issues matter; a set of rules will have to be agreed upon about the social responsibilities associated with the widespread use of Glass (should they be worn while driving is another hot potato). But we're a society that is both highly adaptable and more than capable of working things like this out. It's why Google have adopted the slow-burn pre-release seeding strategy, and if they continue to do a good job in helping to direct the debate and educate a growing army of Glass enthusiasts and ambassadors, all should be well. And, frankly, privacy is not the pivotal issue. Well-marketed and understood Glass is not, nor should become, a problem.

But back to the first question for a moment: is it a solution, and if so to what exactly? Remember this is brand new technology, not an upgrade of something that has come before, and probably the wrong way to look down the telescope at it is to focus on either what Glass is capable of right now (or isn't), or to fixate on the still mildly clumsy hardware (the reported poor battery life is misleading: Glass is not yet running Android 4.3 and so has none of the power-efficiency software improvements Android users are used to. You can be sure that will be fixed in the production model).

> *Glass is unlikely to be the natural delivery device of rich content (Twitter is the type of app that is ideal for Glass).*

with hidden gems while you follow directions using the Field Trip Glassware.

Arguably, Glass will become in time a possible solution to the difficult to pose, but intuitively felt question: how do we seamlessly integrate virtual and data-intensive experiences into our daily lives without becoming more and more socially atomized? We want more social cohesion rather than less, and a great deal of evidence is pointing towards a trend that shows the massive sharing

Walking around the streets of London wearing Glass can make you feel a bit self-conscious like a z-list celebrity, and you can expect to be stopped and interrogated by curious punters – hey, it's still a hot bit of largely unknown and unexplored tech for most people. But after a few days using Glass it starts to become addictive. It's the alerts that do it ... getting an email, text or Tweet popping up in your eye is something you very swiftly grow to appreciate, and being able to reply by voice command means you are never out of touch. And that's what it is all about.

of knowledge, news and non-stop, on-tap commentary on life is leading to the increased collective good. Glass, in its future evolved form and with its selective, dual direction and tuned data feeds, will be the interface that connects each of us to the virtual world – and thus to each other – without the need for boxy, hard-formed, socially interruptive intermediaries. Worn constantly (and it is only a matter of time before sunglass and prescription lens are integrated into the frame), but interacted with in sporadic bursts, will mean good old-fashioned eye-contact will come back into fashion.

Is Glass a problem? In the short term while we reconfigure social interactions to accommodate it, for sure. But the potential, if we adopt the technology, is so enormous that perseverance with it will pay dividends.

And yes, Google Glass Explorer is very much an unfinished idea. But then so was Martin Cooper's brick-like hand-held phone (mobile phone doesn't really describe it). This first iteration of Glass is tantalizingly close to something magical. No doubt it will continue to reside within Google X where brains the size of planets hang out. Mary Lou Jepsen, head of Display Division at the Google X lab, said recently that she and her team are busy working on preparing the next step and they were, 'maybe sleeping three hours a night to bring the technology forward.'

So Google believe in Glass sufficiently to work tirelessly to bring us a version that is small enough, light enough and smart enough to suit every face and mind on the globe. We could very well be witnessing gadget history in the making.

THE ENGLISHMEN IN BEIJING

CHINESE LESSONS

In December 2013, *The Gadget Show* went on air in China. With Chinese presenters, a Chinese production team and a new name: 一键启动 (translation: 'Push The Button Then Start'). In the months leading up to that first TX, a team of four from *The Gadget Show* UK, flew out to lend a hand (and to get a free all-expenses-paid trip to China). These are the 10 lessons they learned out there.

Ewan Keil
Founder of *The Gadget Show* and always up for a free lunch

Colin Byrne
Gadget Show Producer who always dreamed of being a surgeon

Alex Armstead
Gadget Show director, who always dreamed of being Colin

Leigh Nicholls
Part editor, part tourist, all Brummie

THE LESSONS

1 CHINESE RED BULL TASTES NICER THAN BRITISH RED BULL

. . . and it needs to! The Chinese team work minimum twelve-hour days, six or seven days a week. But if they get sleepy, they just lay their heads on the desk and have a snooze for an hour or so. Leigh picked up this working practice quickly – although we had to remind him that it was rude to sleep while someone was actually talking to you.

2 ALWAYS BLESS YOUR CAMERA KIT

On our first shoot for Gadget Show China, no filming could begin until a Buddhist ritual had been observed by the whole crew to bless the camera and give the whole production good luck. It involved a small pot, burning jossticks and bowing to the four points of the compass.

We also have a pre-shoot ritual on the UK show – it's called 'Getting Jason a Starbucks!'

3 GET THE AIR INDEX APP

Air pollution in Beijing is no joke: when it's bad, it burns your throat. We all got addicted to the China Air Quality Index App. We checked it every morning to find out if we'd have to spend the day in face-masks or not.

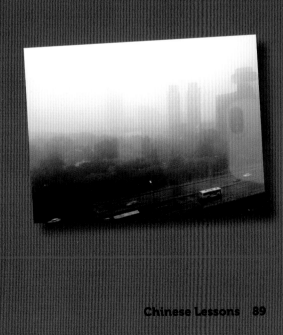

4 MAKE YOUR BOSS GET AN OFFICE CAT AND PAY FOR YOUR LUNCH

Three cats live in the Gadget Show Production Office in Beijing. One of them likes to get involved with whatever is going on. He seems particularly keen on any remote control gadgets — oh and he purrs when he sees a picture of Rachel (but then so does Ewan).

And it's not just the cats that get fed in the office. Everyday at around midday, in the office or on shoots, a man arrives on a motorbike with a big bag containing lunch for everyone. Always with rice and often blow-your-head off spicy, but always delicious. And after two months Alex can almost use chopsticks without swearing! Bless him.

禁止吸烟
No smooking

S TAKE A PHOTOGRAPH OF EVERY SIGN YOU SEE

. . . because when you then have to write an article about your trip to China you can fill up a whole section with the funny signs – and get away with a load less writing without upsetting the editor.

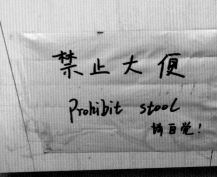

禁止大便
Prohibit stool
请自觉！

6 DON'T LAUGH AT THE CHINESE CREW'S TRACKING VEHICLE . . .

. . . oh, go on then!

7 EVERYTHING TAKES TWICE AS LONG TO SAY IN CHINESE AS IT DOES IN ENGLISH

So, production meetings were interesting . . . and long. Every conversation needed to be translated back and forth between Chinese and English by a team of specially employed translators. And I am convinced that everything takes exactly three times as long to say in Chinese – and when you add in the fact that it takes Ewan three times longer than any normal human being to get to the point, you can see why some production meetings lasted over nine hours!

🄱 BOND WITH YOUR CREW

This is true the world over in television, but it was really important for us to work intimately with the cameramen and local directors to show them the style of shooting required for *The Gadget Show* – very, very different to anything they had done before (Chinese TV is mostly made up of big studio talent shows, big studio dating shows, studio interview shows and drama).

The Gadget Show's Excutive Producer, Ewan Keil hangs out with the Chinese presenters.

9 A FEW SIMPLE RULES OF THE ROAD IN BEIJING

(i) Because congestion is so bad, locals say that to drive anywhere in Beijing takes one-and-a-half hours and they appear to be right.

(ii) Never leave Colin to explain where you're going to a taxi driver – unless you want to film it on your phone and then post it on YouTube a couple of months later (just search for 'Colin in Beijing Taxi' on YouTube – we've posted loads).

(iii) Always carry a picture of your hotel – because no Beijing taxi driver speaks English and many can't read either.

(iv) Don't be scared; the roads are utter mayhem and there appear to be no rules whatsoever, but we never saw a single crash in our whole time there (we did see a girl throw up all down the side of the taxi she was in, though).

⓾ MAKE SURE YOU ARE LOVELY TO EVERYONE

. . . because you will honestly want to go back as soon as possible. China is awesome and the local Gadget Show team in Beijing are some of the loveliest and hardest-working people you could ever meet. The food is great and if you can put up with people constantly asking 'LadyBar?' when you go out at night near your hotel, then the bars and restaurants are fantastic – try the Setzuan fish (it makes your tongue go all numb and fuzzy).

Right that's about it. Now for some more pictures.

666

The Gadget Show's GGGGG Star Performers

Group tests of everyday tech are the lifeblood of *The Gadget Show* and we usually turn to Jon Bentley to bring his planet-sized brain to the task. Over the years only a few prized gadgets have garnered the much sought-after 5G rating. Whatever it is, it has to be pretty special to persuade Jon and the team that it is worthy of the show's highest accolade. Here is a selection of recent top gadget luminaries.

Integral Neon (USB Stick)
Top 5 USB Sticks (Series 16, Episode 11)

When Ortis tested 35 USB sticks with the help of his 'tough' friend Tony, he ran over them with a Land Rover, then washed and tumble-dried them in a launderette, and toasted them on a barbecue. The Integral Neon was the only one that survived.

FACTS

Available in 4 to 64gb capacities and in fluorescent yellow, pink, orange and blue, the lightweight, slim design is both PC and Mac compatible. USB 3.0 compatible and capable of transferring data at up to 5Gbit/s – more than ten times as fast as the 480 Mbit/s top speed of USB 2.0.

Epson Picture Mate 290
(Portable Photo Printer)
Jon test with Chris Packham

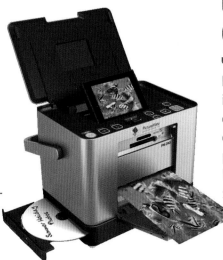

FACTS

The Epson PictureMate 290 is a complete personal photo centre designed for ease of use and to produce high quality photos conveniently at home. With the built in CD-RW drive it will save photos directly to CD without the need of a PC.

Portable versions of bulky things often involve big compromises. But the Epson's astonishing print quality meant you'd choose to use it over your desktop printer for 4 x 6 prints. And those prints have proved durable too – I still have some on the wall at home years later.

Brompton M3L Folding Bike
Jon test with Rebecca Romero (Series 10, Episode 5)

I was an 'all bikes are the same' sort of person until I rode the Brompton. Its nimble handling and light-footed playfulness reminded me of a Lotus Elise – praise indeed. Its folding skills are a splendid bonus!

FACTS

The Brompton is a bike that offers a superb ride, is safe, agile and fast, yet folds easily and quickly into an amazingly compact, portable package. With lightweight Sturmey Archer BSR 3 Speed rear hub gear and durable steel construction, it's also built to last!

DeDietrich DOP895B (Oven)
Jon test with Simon Rimmer (Series 12, Episode 2)

I didn't think an automatic oven that you relied on entirely to set the temperature and declare your dish cooked could possibly work, but this one was a joy to use and produced an excellent roast, comprehensively beating the other two on test.

FACTS

Probably the most stylish and intelligent oven on the planet! The motorised tilting touch-screen control panel guides you through every step of many classic dishes – from roasts to pizzas. The oven will even tell you which shelf to put the food on and how long before it's ready to serve. Energy efficient, incredibly cool and packed with many helpful settings, the DeDietrich DOP895B is a remarkable new take on a familiar home appliance.

Salomon Cosmic 4D GTX (Walking Boots)
Jon test with Brian Blessed (Series 12, Episode 8)

These walking boots were a revelation to me. Light, comfortable and competent straight out of the box. No breaking in or other masochistic rituals required. More walking boots are now designed this way and that's a very good thing for walkers' feet.

FACTS

Tough and comfortable, the Cosmic 4D GTX is a state-of-the-art hill and mountain boot for fast, light travel on a variety of terrain. The Salomon Cosmic 4D GTX® draws on Salomon's trail running heritage and uses the latest materials to make a boot with sufficient stiffness for mountain walking and backpacking with a light and agile trainer-like feel.

Sonos ZonePlayer S5
(Wireless Music Player)
Jon test with Mark Goodyear
(Series 13, Episode 4)

Some companies specialise in a market niche and do it most excellently. Sonos have consistently produced brilliant wireless music systems for the home and their ZonePlayer S5 breezed through this test impressing broadcaster Mark Goodyear as well as me.

FACTS

The original, big sound and wireless speaker. The ZonePlayer S5 uses five state-of-the-art digital amplifiers individually coupled with five speakers to deliver superior sound to any room. A subwoofer produces powerful bass while two mid-range drivers and two tweeters fill out the sound. Different music sources from one app combined with complete multi-room capabilities mean you can listen to your entire world of music in every room.

Stûv 30 Wood Burning Stove
Jon test (Series 14, Episode 18)

Wood burning stoves don't have to be nostalgic exercises in mediaeval rustic ironwork. Stûv's stove looked bang up to date, and was ultra clean and impressively efficient. And when you tuned up the heat it was scorchingly powerful too.

FACTS

Function and form: the revolutionary Stûv 30 ensures you always have a full view of the fire with its built-in turning plate that means the stove can be rotated in any direction. The system allows rotation up to 360° or can be custom limited during installation.

GoPro HD Hero2 (Video Cam)
Jon test (Series 13, Episode 2)

GoPros have changed the way we see the world. Whatever your challenge or sport, you can now record your participation in HD from virtually any angle. Who would have thought a video camera could be so liberating and powerful a communication tool? And such great fun too!

FACTS

The GoPro HD Hero2 has proved itself to be the ultimate Gadget-Show-proof action stills and video camera of the last decade. Designed to be mounted on pretty much anything – head, chest, surfboard, car and even the odd balloon flight to the edge of space! It's been up-rated since this test and just gets better with each incarnation. Capable of 60 frames per second in 1080p (1920 x 1089) with the f/2.8 fixed focus lens providing up to 170 degree field-of-view to capture all the action.

HTC One (Smart Phone)
Jon test in Amsterdam (Series 17, Episode 4)

Arguably the world's first truly desirable Android phone, the HTC One is a large yet manageable sleekly tactile slice of one-piece aluminium gorgeousness. It shot to the top of the world's most desirable smart phone list. An instant classic.

FACTS

The HTC One combines stunning design, a superlative screen and immense processing power. It has a full HD screen crammed into 4.7-inches with 468ppi delivering incredible sharpness. Added to which there is a CPU and RAM combo that set the benchmark other brands have had to follow. It's loaded with 32GB of storage and Bluetooth, Wi-Fi and 3G and 4G connectivity.

XBOX 360

FACTS

You are the controller! Kinect offers a revolutionary way to play, share, communicate and interact with the Xbox 360. With Kinect, your body becomes the controller thanks to the Kinect Sensor bar, a high-tech gizmo that can track your movements, recognise your voice and even scan objects into games for a completely immersive experience.

Xbox Kinect
Jon test (Series 14, Episode 19)

Creating a new gesture-control interface was a brilliant achievement by Microsoft in 2011 – a development that's sure to inspire future technology beyond the Xbox, and encourage more widespread use of getting what you want by waving your hands in the air.

Dyson DC25
Jon test (Series 11, Episode 2)

With its innovative range of vacuum cleaners, the first of which appeared back in 1993, Dyson transformed a boring household chore into a pleasurable experience. With its ball technology, the DC25 proved satisfyingly steerable, while its great suction and fine anti-allergy filter contributed to a brilliant cleaning performance.

FACTS

The DC25 Multi Floor is engineered to clean every floor type and has a Quick-Draw Telescope Reach wand for stairs and hard-to-reach places. Conventional upright vacuum cleaners ride on fixed wheels, but can make you work a lot harder than you need to! Dyson Ball vacuums ride on a ball, so you can easily change direction with just a turn of the wrist.

Samsung UE55D8000 TV
Jon test with Simon King (Series 16, Episode 5)

Samsung's super stylish, ultra slim 55 inch TVs enjoyed a good innings at the top of the telly tree. This set not only beat its rivals on the day, it also converted a sceptical Simon King, one of Britain's best wildlife cameramen, to the joys of 3D.

Cewe Photo Book
Jon test (Series 17, Episode 9)

We measure our lives in memories, and photobooks are an affordable and practical way of cherishing the best. In our tests the easy-to-create, visually impressive Cewe beat its competitors to produce a lasting memory of Rachel's first series on The Gadget Show.

FACTS

The free downloadable software makes designing your photobook relatively straightforward and there is no doubt the quality of the final result is exceptional. The lie-flat photobook has a hinged binding that provides additional display space and means you can view the images from most angles, safe from that annoying glare of overhead lights.

FACTS

The UE55D8000 really stands out from the crowd with its almost bezel-free design that is truly immersive when watching 3D broadcasts. It was Samsung's flagship 2011 model and came fully loaded with a host of innovative features. The 'Smart Hub' – an on-screen menu system accessed via a dedicated button on the remote control – and a beautifully designed and high resolution home screen took you to any video, photograph or music file stored on a USB stick (via one of the three USB ports). It had two built-in HD tuners (Freesat and Freeview) and was wifi-enabled, meaning a networked computer could be accessed or the Internet searched via the built-in web browser. In 2011 TVs didn't get much smarter than this.

Canon EOS 700D (DSLR)
Jon test (Series 18, Episode 1)

Digital SLRs have made top quality film making very affordable. A few years ago you'd need tens of thousands of pounds worth of kit to match the quality of Canon's entry level 700D. And the stills I took during my day with The Sealed Knot weren't bad either.

FACTS

Step into DSLR photography and let your creativity grow. The Canon EOS 700D will produce superb photos and video with an 18-megapixel sensor and has an easy to use Vari-angle Clear View LCD II Touch screen. It can create high-quality low-noise images that are packed with detail.

COMPUTING
ULTIMATE GADGET FACE-OFFS

In 1965 Intel's co-founder Gordon E. Moore predicted that the number of circuits included in, and therefore the processing power of, computers would double every eighteen months for 'at least the next ten years'. Forty-five years later we are still living in a world that is subject to 'Moore's Law'.

While the overall clock speed of processers for the domestic market seems to have peaked at 3.5-4 GHz, the size of the circuitry used is forever shrinking, leading to miniaturized and increasingly powerful chips being employed in more and more applications and devices.

When you compare the computing power used during the Apollo 11 Moon landing at 2.048 'MHz' (roughly the processing power you will find in a basic pocket calculator) with that of the PC this is being written on (it has a 3.20 'GHz' Quad Core processor) you are a witness to Moore's law in action over time.

Here we put the innovators of computer tech, its peripherals and related software face-to-face, screen-to-screen.

NETSCAPE

Success: 4
Following its release Netscape soon became the dominant web browser of the 1990s.

Lifetime: 2
By 2002 it had pretty much disappeared as Microsoft's Internet Explorer (IE) achieved its dominant position helped by it being bundled as part of the Windows 95 operating system.

Innovations: 4
The first popular web browser.

Fun Facts
In a *PC World* column published in 2007, the original Netscape Navigator was considered the 'best tech product of all time' – owing to its initial impact on the widening public use of the Internet.

RASPBERRY PI

Success: 5
Sales commenced 29 February 2012 and by September approximately 500,000 boards had been sold.

Lifetime: 4
Although relatively new, the Raspberry Pi already has a committed education-focused community built around it – and looks set to expand.

Innovations: 5
It's a basic, affordable computer that allows people, especially kids, to learn about programming. Very cheap and flexible. What's not to like?

Fun Facts
The UK's two licensed manufacturers, who sell Raspberry Pi's online, both had their websites collapse thanks to exceptionally heavy web traffic on the announcement of the release date. At one point a webmaster pleaded, 'Guys, can you please stop hitting F5 on our website quite so often? You're bringing the server to its knees.'

GOOGLE CHROME

Success: 4
Within four years of its launch, Chrome has become the most widely used web browser in the world.

Lifetime: 4
December 2008 to now.

Innovations: 4
The first browser to concentrate on working with web applications rather than just web sites.

Fun Facts
For six years Google's Eric Schmidt opposed the internal development of an independent web browser, but when Chrome's developers showed him a demonstration Schmidt admitted, 'It was so good that it essentially forced me to change my mind.'

ZX SPECTRUM

Success: 4
Released in April 1982, demand far exceeded supply. By July there was already a backlog of orders of 30,000 units that soon turned into 40,000.

Lifetime: 5
Since the Spectrum's launch 24,000 software titles have been released. Over one hundred new titles were released as recently as 2012.

Innovations: 4
One of the first mainstream home computers in the UK that ultimately led to a boom in companies producing software and hardware for the machine.

Fun Facts
A number of today's leading developers began their careers on the ZX Spectrum including Tim and Chris Stamper.

IPHONE

Success: 5

In excess of 250 million units sold in the five years since launch.

Lifetime: 4

29 June 2007 to now.

Innovations: 5

The first smart phone. The iPhone was the first consumer piece of hand-held tech that allowed the user to use applications (programmes in earlier computer-speak) and that changed everything. The (smart) phone was no longer a simple communication device or an object purely designed for business use, becoming instead something desirable and life enhancing.

Fun Facts

iPhone development began back in 2004 when Apple gathered together a 1000-strong team of employees to work on the highly confidential 'Project Purple'.

APPLE MAC

Success: 4

At the end of 2012 the Apple Mac attracted forty-five percent of the profits in the personal computer market.

Lifetime: 4

1984 to now.

Innovations: 5

Apple Macs were the first computers to ship with mice among many other innovations (including the culture-shifting Graphic User Interface).

Fun Facts

Hard to believe now, but when the first Macs were shipped, they had just 129kb of memory and no hard drives . . .

TOUCHSCREEN

Success: 4

By 2010 the majority of smart phones had touch screens.

Lifetime: 4

The capacitive touch screen was first introduced on the LG Prada in May 2007. By 2012 Windows had integrated it within their main operating system.

Innovations: 4

It feels so natural and intuitive, allowing the user a tactile simple interface with machinery. You simply poke or swipe at the thing you want to respond.

Fun Facts

Some of the first touch screens worked by utilizing infrared lights to detect disruption in the pattern of light.

PSION

Success: 4

When Psion announced they were to stop making new consumer handheld computers they had sold between five and six million PDA units since it first entered the sector in the 1980s.

Lifetime: 4

1984 to 2001.

Innovations: 4

The world's first handheld computer; it was capable of collecting, storing and processing data in the field, and then synchronizing with a PC.

Fun Facts

The name Psion is derived from Potter Scientific Investments after the company's founder David Potter. Marks & Spencer were always a big fan of Psion using them to keep track of prices at their checkouts.

THE PC

Success: 5

There are over one billion PCs in the world; globally it simply dominates as the main computer on desks and in homes.

Lifetime: 5

We have lived with the PC as part of our work and home lives for more than thirty years.

Innovations: 4

The PC was the world's first practical computer for the masses.

Fun Facts

The Commodore PET (short for Personal Electronic Transactor) was launched in 1977 with either 4kb or a premium 8kb of RAM and several magazines have designated it as 'the world's first personal computer'.

KEYBOARD AND MOUSE

Success: 5

The keyboard remains the single most common method of computer input since its inception more than forty years ago.

Lifetime: 4

We have had computer keyboards for more than forty-four years and it has been twenty-five years since the mouse first took up residence on our desks.

Innovations: 4

In 1984 Apple Mac computers were the first to use mice.

Fun Facts

Doug Engelbart (who died in 2013) was the inventor of the computer mouse. His ideas were way ahead of their time in an era when a computer took up an entire room and data input meant feeding huge machines with punch cards. He first demonstrated the mouse during a presentation that became known as the 'mother of all demos' in 1968 in San Francisco.

WHAT DO YOU KNOW ABOUT?
Rachel Riley

Rachel has quickly established herself in *The Gadget Show* studio forging a lively partnership with Jason (well, you do need a lot of energy to keep up with him). Oxford graduate Rachel made her television debut in 2009 after her mum filled in the application form for Carol Vorderman's replacement on *Countdown*!

Here are five things you might not know about Rachel ...

1. Over the years I have had loads of exotic pets including poison dart frogs (which are an aposematic organism, i.e. their bright colouration warns potential predators that they are unpalatable!), snakes, lizards, tortoises and a giant African bullfrog.

2. I've been on loads of television quizzes on my own or in teams including The Chase, Mastermind, Million Pound Drop. Along the way I have managed to win over £200k for charity!

3. When I was little my favourite toy was a computer that had numbers games to play with, and I won my first prize for mathematics at the age of six.

4. At the University of Oxford I played centre-midfield for Oriel College ladies football team. We won our league!

5. When I was nine I won a competition on TCC (The Children's Channel) and my prize was a yellow Game Boy and a signed picture of Ortis Deley!

The Console

The Gadget Show has always reported on the cutting edge of console gaming technology. Not surprising really when you consider that Jason Bradbury is an ex-gaming journalist and we have lots of input from industry expert Julia Harding and gaming's YouTube superstars Ali A and KSI. Our intention has been to keep you right up to date with news of the latest hardware and software releases. We thrive on this technology just as much as you do at home!

From its humble beginnings with arcade game *Pong*, gaming has grown into one of the biggest entertainment industries on the planet: *Grand Theft Auto V* racked up $1 billion in worldwide retail sales in three days after its release in September 2013, proving, if proof was needed, the sheer might of the industry. Which makes it all the more interesting for industry observers as we enter the new era of eighth-generation

Gaming Genome

An Examination Of The DNA Of The **PlayStation 4** And **Xbox One**

gaming consoles. Two of the biggest players, Microsoft and Sony, have released their latest generation consoles almost simultaneously, and both machines have very similar specs. So what do these new powerhouses of gaming offer users above and beyond the previous generation? And what has actually led both companies to this point in gaming history?

Over the next few pages we hope to offer you an insight into the twelve-year battle between the two console giants: the tech background, the history and evolution of the respective bloodlines of both consoles and their shared gaming DNA. Make no mistake, it is a console war, and, perhaps more than ever, which company ultimately wins out will be decided not on price or processing power, but on the latest generation of software and gaming titles that will showcase the eighth-generation consoles' graphics-handling and interactive capabilities the best.

THE PLAYSTATION GENERATION. IT ALL STARTED ON A WEDNESDAY

On 27 October 1993, the world learnt officially about Sony's concept of a 32-bit next generation gaming console. What was unusual, however, was that the announcement wasn't made via the usual slick, multi-media, full-on press event, but instead through a simply-worded press release, distributed that Wednesday morning. In it Sony Corp informed the global press that it was:

'Working toward marketing the new home-use game system domestically by the end of 1994, and overseas within 1995, priced competitively.'

In another press announcement a week later they went a step further:

'The next-generation games machine is expected to offer high-speed, simultaneous movement of characters and high-quality backgrounds together with powerful 3D computer graphics.'

Omitted at this point was any mention of a release date or what the console was going to be called. What came next, however, launched a console bloodline that has, twenty years later, entered its fourth tech evolution. The launch established Sony as both the originators and true innovators of the burgeoning gaming industry. Nearly one month after the first announcement, Sony revealed the working title – the PlayStation X (PSX). The name PlayStation was actually derived from an earlier Sony project to design a CD-ROM drive for the Nintendo SNES, but as a mark of the eventual success of the console concept, the term 'PlayStation Generation' would become synonymous with young people of the nineties and early noughties and, in particular, young game players – regardless of the

SPECS: PLAYSTATION

CPU: R3000A 32bit RISC chip @33.8mhz **GPU:** R800A (33mhz); 16.47 million colours; Sprite/BG drawing; Adjustable frame buffer; No line restriction; Unlimited CLUTs (Colour Look-Up Tables) 4,000 8 x 8 pixel sprites with individual scaling and rotation; Simultaneous backgrounds (Parallax scrolling) 620,000 polygons/sec **Resolution:** 640 x 480, **Memory:** 2MB System RAM, 1MB Video RAM

actual gaming platform they used. The popular press used the term to suggest that many young gamers had somehow 'lost their childhoods' because of video games, but the reality was that they had had, for the most part, their lives enriched by the injection into it of modern tech, something there would be no going back from.

SPECS: PLAYSTATION 2

CPU: 300 MHz MIPS 'Emotion Engine' **GPU:** 147 MHz Graphics Synthesizer, fill rate 2.352 gigapixel/sec; Capable of multi-pass rendering; Able to deliver enhanced shader graphics and other enhanced graphical effects **Resolution:** 640 x 576, **Memory:** 32 MB of RDRAM, 4 MB eDRAM

SIX YEARS LATER . . .

Surprisingly, it would be another six years before Sony returned to the by now increasingly competitive console market, this time with the 64-bit PlayStation 2. On 2 March 1999, the new machine was introduced to the world at the less-than-imaginatively titled 'PlayStation Meeting 99'. At first known as 'The Next Generation PlayStation', the new machine promised gamers an integrated DVD player, a powerful 64-bit Emotion Engine CPU and the ability to handle twenty million polygons per second, compared to the 620,000 of the original PSX. In stark contrast to the drip-feed PSX debut, on 13 September 1999, 900 journalists and members of the public were invited to a lavish Tokyo event where PlayStation 2 was officially unveiled along with launch title games like *Tekken Tag Tournament* used to show off its new graphics-handling prowess.

LOS ANGELES E3 USED AS A LAUNCH PAD FOR THE THIRD WAVE

On 16 May 2005, following yet another six-year development cycle, Sony used its pre-Electronic Entertainment Expo (E3) press conference in Los Angeles to announce the new PlayStation 3 console. There had been months of industry speculation about its possible specs, but it took just a few minutes for the company to put all the rumours to bed and stun the gaming and tech press with the new console's ground-breaking multi-cell processor and standard

Blu-ray drive. The sleek new design, that would perfectly complement any tech-friendly home, when coupled with a major leap forward in its graphics handling ability – as demonstrated by launch titles *Metal Gear Solid 4* and *Final Fantasy VII* – meant that Sony had significantly raised the bar of high definition graphic-intense gaming.

NEW YORK WAS THE PLACE TO BE

On 20 February 2013, eight years after the launch of the PS3, Sony finally revealed the detail of its latest and hotly anticipated console at a dedicated PS4 event in New York. With over 1,000 journalists in attendance, and live footage streamed to countless eager fans all over the world, Sony made a potent presentation of the new console's features. The company was certain that this was the future of gaming – a machine with a powerful architecture, a new controller and fresh digital distribution concepts at its heart. What

the connected cables. The black-themed colour, along with the combined matt and gloss finish, adds a really sophisticated look, and as a next generation entertainment system, it fits in perfectly wherever it is placed in the home. The PS4 peripherals include a stand-alone wireless controller, the Dualshock 4, and PlayStation camera. Both accessories are 'jet black', so, all in all, the whole set up is seriously *uber-cool* . . .

SPECS: PLAYSTATION 4

CPU: Single-chip custom processor: 8 Core x86-64 AMD 'Jaguar' **GPU:** 1.84 TFLOPS; AMD next-generation Radeon based graphics engine; Blu-ray D/DVD Drive (Read Only); 2 Super-Speed USB 3.0 2; 1 x AUX port **Networking:** Ethernet (10BASE-T, 100BASE-TX, 1000BASE-T) x 1 IEEE 802.11 b/g/n; Bluetooth 2.1 (EDR); **AV Output:** HDMI out port; Digital Output (Optical) port **Resolution:** 1080P HD and support for 4K output, but only for photos and videos, not games. **Memory:** GDDR5 8GB; Storage size 500GB hard disk drive

SPECS: PLAYSTATION 3

CPU: 6-Core Cell Broadband Engine (3.2 GHz Power Architecture) **GPU:** 550 MHz RSX 'Reality Synthesizer' (based on Nvidia G70 architecture); 1.8 T-FLOPS floating point performance **Resolution:** 1920 x 1800 Full HD **Memory:** 256MB XDR Main RAM; 256MB GDDR3 VRAM

was perhaps surprising, given Sony's demonstrable confidence in the new PS4, was the absence of even a prototype unit on show! One of the core messages of the presentation, however, was the repeated mantra: '*It's a console built by gamers for gamers.*'

The actual design of the PS4 made its public debut on 10 June 2013 at a PlayStation press conference that kicked off the year's E3. The internal design architecture of the PS4 system, from its optical drive and power supply unit to the cooling mechanism, had been pared down to keep the body as slim and light as possible and to dramatically enhance the flexibility of the design.

The PS4 features a simple but modern design accentuated by its linear flat form. The surface of the PS4 body is subdivided into four sections as if four blocks are stacked together, making one object, with disc slot, buttons, power indicator and vent placed within the slim interspaces created between the sections. The power indicator on the top of the body glows in PlayStation blue when the device is powered up. The front and back of the unity are slightly angled giving easy access to the power button and disk slot when placed either horizontally or vertically and helping to conceal

The PlayStation 4 has a whole host of exclusive titles, and multi-media capabilities, and from the outset there have been zero restrictions on second-hand games and compulsory online connection.

The Real Generation X

A DIRECT X BOX

Some five years after the release of Sony's PSX, four Microsoft employees apparently pitched an idea to head honcho, Bill Gates – for a gaming console. It was initially called the 'Direct X Box' and was basically a robust PC-based gaming rig built into an easily identifiable console case. The system ran on a version of Windows 2000, making it easy for existing PC software developers to work with the console's architecture that, at the time, boasted almost twice the processing power of newly released and rival PlayStation 2. After a further two years development, Microsoft was ready to go public with the console. Bill Gates famously whetted the appetite of the gaming world at the Game Developers Conference in 2000. He made a point of emphasising the high specs of the machine and its potential for realising on screen its eventual breath-taking graphics. Both offline and online play would be possible and, because the console came equipped with internal data storage, users would be able to download memory-hungry content for both games and other media. Once Bill Gates had wrapped up his presentation it was clear that Microsoft had secured the gaming world's attention and from now on were going to be the undaunted new kids on the gaming block.

THE HALO FACTOR

Just months later at E3, one of the Xbox development team met with Bungie Studios. The studio had developed a shooter game called *Halo: Combat Evolved*, and the consequence of the fateful meeting is enshrined in gaming history. Microsoft bought Bungie Studios for US$30 million and, after further development and tweaks, *Halo* became the game that would showcase the new console's amazing graphics-handling abilities. In so doing *Halo* would become one of the most iconic game titles of all time.

Initially exclusive to Xbox, *Halo's* sales shattered all previous records, selling more than one million copies during the first six months of its release.

XBOX 'Я' US

In 2001 Bill Gates, with a little help from Dwayne 'The Rock' Johnson, revealed the final design of the Xbox at CES in Las Vegas. As well as introducing technical demos, Gates told the gathered press that the official US release date would be 14 November that year. The Times Square branch of Toys 'R' Us would be the official launch site with Bill Gates himself opening the proceedings. Industry-wide comments about the console's final design and its (high) US$299 price tag were quickly silenced as over one million Xbox consoles were sold in just three weeks.

Xbox was released in Japan on 22 February 2002 and finally in Europe on 14 March. Both markets were more reluctant adopters, failing to generate anywhere near the sales witnessed in the US. On 18 April 2002, Microsoft made the move that boosted its global sales of the Xbox by slashing the price to US$199. Immediately units started to fly off the shelves. But with a reportedly high manufacturing cost, it was rumoured that the Xbox was being sold at, or close to, a financial loss. Perhaps the business plan was based on Microsoft's traditional (gaming in this instance) software licensing model, but whatever the truth of the matter, the Xbox was finally selling in the numbers originally anticipated by Microsoft, even managing to outsell the newly released Nintendo GameCub. From now on it would be impossible to argue with the fact that Microsoft had established itself as a major force in the console market.

XBOX'S LIVE BIRTHDAY PRESENT

A year after the arrival of the Xbox, Microsoft introduced Xbox Live. The online gaming network began beta testing

in August 2002 and the system was launched to the wider public in November with the Xbox Live Starter Kit. This added an online multiplayer gaming experience, hitherto something unheard of in the console gaming world. Xbox Live also made it possible to download new content and took multiplayer interaction to a whole new level. Over 150,000 users subscribed to the service in just seven days and thereafter the numbers kept doubling. While the Xbox user base continued to grow, with the help of Xbox Live, sales of the console itself eventually started to diminish. By 2003 Microsoft had already begun designing the next generation console – codenamed 'Xenon'.

ONE FINAL REALLY BIG ACHIEVEMENT

With Microsoft continuing to add new features to Xbox Live – and its soon to be announced successor – the company's ambition to make their games console a genuine media hub moved a step closer with Xbox Live Arcade that

launched in November 2004. While the Xbox was in, what industry observers decided, terminal decline (regardless of reports that information about its replacement was just over the horizon), the launch of *Halo 2* proved that rumours of Xbox's imminent demise were much exaggerated. Over two and a half million copies of *Halo 2* sold in the first twenty-four hours. The game's sequels have made over US$125 million in sales globally, making it the most successful launch of any entertainment product (including Hollywood blockbusters) in history. That said, the financial success of *Halo 2* made it a distinctly strange commmercial bedfellow with the Xbox console that was still reported to be costing Microsoft significantly more to produce than it generated in revenues.

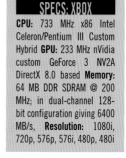

SPECS: XBOX

CPU: 733 MHz x86 Intel Celeron/Pentium III Custom Hybrid **GPU:** 233 MHz nVidia custom GeForce 3 NV2A DirectX 8.0 based **Memory:** 64 MB DDR SDRAM @ 200 MHz; in dual-channel 128-bit configuration giving 6400 MB/s, **Resolution:** 1080i, 720p, 576p, 576i, 480p, 480i

INTRODUCING THE NEXT GENERATION

On 12 May 2005, Microsoft unveiled the Xbox 360. The Xbox 360 was due to ship in November that year, nearly twelve months before the PlayStation 3 and Nintendo Wii were scheduled to arrive. It would allow the console a foothold in the market, and for the first time positioned Microsoft as a now dominant force in the industry. Later that year production of the Xbox finished with total sales climbing to twenty-four million units; less than half of the original fifty million sales originally predicted by Microsoft. In contrast the PlayStation 2 had sold over one hundred and fifty million units worldwide.

The Xbox 360 was released in the US on the 22 November 2005 and in Europe and Japan four weeks later. It instantly sold out everywhere except Japan, and in just one month, more than one and a half million consoles had passed into the hands of eager gamers. But worryingly in the rush to get the 360 to consumers, concerns about the software, and more crucially, the hardware itself started to emerge and then become widely reported.

THE RED RING OF DEATH

A tell tale red ring of light around the console's power button indicated that the device had developed a 'fatal error'. Quickly the Internet was awash with 360 users experiencing this highly unsatisfactory state of affairs and pleading for help. Microsoft was forced to repair or replace an undisclosed number of consoles, and to extend the 360's warranty period because of the amount of downtime experienced by many gamers. Undeterred by this initial and soon forgiven problem, gamers readily fell in love with the 360, and within four years Microsoft had sold more than thirty-nine million units.

360 GETS 'KINECTED'

At E3 on June 2009, Microsoft demonstrated the hands-free motion control system codenamed 'Project Natal'. This peripheral camera and motion sensor device meant 360 owners could now interact with games in a way they

had only dreamed of before. No longer would you need a controller: the motion sensor would allow gamers to become part of the action, controlling games with gesture and body posture alone. Named Kinect, during the following year's E3 Microsoft announced the unit would arrive on shelves worldwide in November. Just twelve months after Kinect was unveiled Microsoft announced the 360 would be getting a facelift. It would be slimmed down, benefit from upgraded internal specs, have additional USB ports and built-in Wi-Fi. This new 'S' model was released on 18 June 2010.

SPECS: XBOX 360

CPU: 3.2 GHz IBM PowerPC tri-core codenamed 'Xenon' **GPU:** 500 MHz ATi custom Radeon X1800 DirectX 9.0c based GPU codenamed 'Xenos' **Memory:** 512 MB of GDDR3 RAM @ 700 MHz 22.4 GB/s; 10 MB EDRAM GPU frame buffer memory **Resolution:** 1080p, 1080i, 720p, 576p, 576i, 480p, 480i, also various monitor resolutions available via VGA and HDMI/DVI

THE XBOX 720

Within a year the industry rumour mill had started up again and was alive with intense media speculation about the likelihood of a new console. Hungry journalists even decided to give it an unofficial name – the 'Xbox 720'. The buoyant sales of the 360 continued to prove Microsoft had

XBOX ONE

SPECS: XBOX ONE

CPU: 1.75 GHz AMD x86-64 eight-core CPU codenamed 'Jaguar' **GPU:** 853 MHz AMD Radeon HD 7000 series DirectX 11.1 based GPU codenamed 'Durango' **Memory:** 8 GB of DDR3 RAM @ 2133 MHz 68.3 GB/s; 32 MB ESRAM GPU frame buffer memory **Resolution:** 4K, 2160p, 1080p, 1080i, 720p

secured a place at the top table of the gaming console industry, and in September 2012 Microsoft announced it had passed the seventy million units sales mark. By December that number had jumped to over seventy five million and although the sales of the 360 actually fell well behind Nintendo's Wii, that had sold one hundred million since launch, the Xbox had outsold its Sony rival, the PS3, by an estimated five million. A month later Microsoft launched Xbox Music and Xbox Video. These new media services offered a clean Apple-style integration across all devices running Windows software. But where would Microsoft go from here?

Well just nine months later on 21 May 2013 at its Redmond, Washington headquarters, the new corporate gaming vision and console were unveiled. The company was keen to demonstrate how Xbox One would put the user at the centre of a multi media experience, encompassing gaming, television, movies, music, sports and Skype. It stressed how integrated the whole experience would be

and that the Xbox One represented the next level in user interaction with an entertainment device.

Don Mattrick, president, Interactive Entertainment Business at Microsoft said:

'Xbox One is designed to deliver a whole new generation of blockbuster games, television and entertainment in a powerful, all-in-one device.'

'Our unique, modern architecture brings simplicity to the living room and, for the first time ever, the ability to instantly switch across your games and entertainment.'

It is rumoured that the Xbox One will have the longest period on sale of any Microsoft product to date, with a predicted ten-year lifecycle until the next generation is released. Whether this is true is questionable given the accelerating rate of development of all manner of tech, but the specs of the console are impressive and point to it not becoming obsolete too quickly!

The impressive processing power means that graphics of the Xbox One are a huge step up from its previous incarnation.

COMPARING THE TWO. IT IS POWER TO THE PEOPLE

So will it be Sony's PlayStation 4 or Microsoft's Xbox One that joins your household? It is a tough choice for any gamer, and brand loyalties aside, now is by far the most exciting generation of console-based gaming since the launch of the original PlayStation. On paper the PS4 is, graphics-handlingwise, the more powerful machine, but sheer processing grunt isn't always the most important feature. Gameplay, connectivity and the user's own immersive experience are far more pertinent. Just look at what Nintendo did to the market with the appearance of the original Wii console: the game play came first and by innovatively introducing motion sensors and gamepads, added a visceral element of interaction to the user's experience. It changed the way games are played for the foreseeable future. What Sony and Microsoft have done is to take the interface between the gamer and the game to an all new level of sophistication. Microsoft's new Kinect II sensor and Sony's Eye Camera are far more sensitive and accurate than their predecessors; the Sony camera alone is capable of a video frame rate of 1280 x 800 pixel @ 60fps, with a field of view of 85°, and the unit has a very sensitive four channel microphone array.

The Xbox One's Kinect II will run at a resolution of 1080p and is three times as sensitive as its predecessor. It boasts an improved field of view allowing up to six people to be sensed and seen on screen at once! It also has the very cool ability to sense infrared light, which means it will now detect and track a player perfectly in a pitch black room. As far as software is concerned, the improved skeletal mapping is far more accurate, and will track individual hand motions and nuances – such as a player shrugging their shoulders. And it's not just skeletal: Microsoft have programmed in a muscle tracker that will, by accurately tracking speed of movement, calculate which parts of the game player's body has pressure exerted on it, calculate where weight is being distributed and how much power has gone into each gesture.

PROCESSING POWER AND GRAPHICAL MIGHT

AMD has made the chips used by both companies and each uses 'system on a chip' (SoC) that combines the CPU and GPU. The processor of choice is an x86 64-bit chip with 8 'Jaguar' cores, but there is a marked difference between the two consoles. Sony has a CPU with eighteen 'compute units' built into the chip, but at the time of writing hadn't officially confirmed the clock speed. Speculation is that it will be either 1.6, 1.9 or 2.1 GHz, which when combined with 8GB of ultra-fast GDDR5 RAM will provide real processing crunch, while the Xbox One has a CPU clocked at 1.75GHz with twelve 'compute units' and 8GB of DDR3 RAM. Sony's GPU is capable of 1.84 TFLOPS of throughput against the 1.33 TFLOPS of the Xbox One. Now on paper this may seem like a big power advantage for Sony, but it is not the full story. The Xbox One's GPU clock speed was originally set to be 800MHz, but was revised to 853MHz. This increase may seem tiny, but will allow extra overhead for functions like model smoothing anti-aliasing and other fine-touch graphics effects. Alongside the DDR3 ram, the console also has a further boost through its 32MB of embedded high bandwidth ESRAM. In reality we may never notice the difference between the two consoles in use, especially in the first wave of titles, as developers themselves get used to the architecture and programming.

MULTI-SCREEN ACTION

With certain game titles both consoles offer the ability to use a second device as an additional screen by synching with phones and tablets. Proprietary apps will allow cross-platform connectivity via Android, compatible Windows 8 and touch screen devices, and of course the obligatory iOS support for iPads and iPhones. Just think of the possibilities! For example when playing *Battlefield 4,* a second-screen will allow access to maps, stats, and server browsing. Expect this sort of feature to be embedded in many key titles. Microsoft call it SmartGlass, and it enables you to not only interact with games in real-time but also navigate the Xbox dashboard, pause and rewind films, swipe, pinch, and tap to surf the Web on your TV, and even use your device's keyboard for easy text input during gaming or for using the search functions on the console.

Sony's own PlayStation app can also be used as a second screen in PS4 games, either for mapping or as an interactive additional controller. Interestingly, you can also monitor all your friend's PS4 activity and even send direct messages to them through the app while gaming. The app enables you to initiate multiplayer matches and view gaming information and live broadcasts. You will even be able to use your smart phone as a stand-alone controller for certain PS4 games. Support for this on each console will be driven by third-party developers, but Sony and Microsoft's own in-house teams will, we are sure, utilise these interesting features in some unique and fascinating ways. One big plus for the PS4 is that it can also be used as a server via the app to stream mobile games directly to your device. You won't actually need to download the complete game to your device – instead content will be streamed in real-time as you play!

STORAGE, MEDIA AND CONNECTIVITY

As for built-in storage, it's neck and neck as both consoles possess a 500GB hard drive, although obviously the amount available to the user will be slightly less because of the space needed for essential system software. Looking at the types of media each machine can access, both Xbox One and PS4 have a Blu-ray/DVD drive on board as standard, something that until now always raised the PS3 above the DVD-only capable Xbox 360. Admittedly the latter did get an add-on HD-DVD drive, but as the HD disk format war

has been well and truly won by Blu-ray, the HD-DVD is now well and truly extinct.

When it comes to connectivity it is pretty close, with the PS4 and Xbox One both offering Wi-Fi, Ethernet, optical out and USB 3.0. The PS4 offers Bluetooth 2.1 while the Xbox One does not. Both, as expected, have HDMI ports (supporting the new higher definition 4K output). The Xbox One does have an ace up its sleeve though, in having an HDMI in-port for connecting other HD sources and allowing their control from the Xbox One dashboard.

BETTER FEEDBACK, RESPONSE AND ENTERTAINMENT

The new Xbox controller has much the same design as previous generations, but has incorporated into it a number of subtle changes. A better D-pad and thumb-stick design improve ergonomics and user input, while new impulse triggers offer faster response and better feedback by rumbling too. In contrast the PS4 DualShock 4 controller is very different to its predecessor. A capacitive track-pad is located in the middle of the unit that also has a fully integrated 'move' motion control built in, giving players extra 'immersion-in-the-game' experience. There will also be the option to use your existing Sony PS Vita hand-held console as an additional controller with certain titles!

From the outset these two consoles were designed to be all-round entertainment devices, although the initial pre-launch suggestion that they were going to require a permanent online connection to function proved to be a major bone of contention. Microsoft had announced that the Xbox One would require 'validating' online at least once every twenty-four hours, but backtracked after the predictable outcry, saying:

'An Internet connection will not be required to play offline Xbox One games. After a one-time system set-up with a new Xbox One, you can play any disk based game without ever connecting online again.'

Sony's supposed always-on Internet connection was never, in fact, an actual issue. Shuhei Yoshida, president of Sony Worldwide Studios, spelt it out:

'Did we consider it? No, we didn't consider it. It makes sense for people to have Internet

connections to play online games, but for offline games there are many countries that we saw do not really have robust Internet.'

THE PRE-OWNED DEBACLE

Microsoft seemed determined to alienate its long-time Xbox fans when, in the run up to the release of the Xbox One, rumours began circulating that in-built restrictions would limit owners' ability to lend games to fellow players. The borrower would have to have been on the lender's friends list for more than thirty days, and even then the loan could be made just once. In another U-turn Microsoft had to stand up in front of the world's media to put Xbox fans' minds to rest:

'Trade-in, lend, resell, gift, and rent disc based games just like you do today. There will be no limitations to using and sharing games, things will work just as they do on Xbox 360.'

LACK OF BACKWARD COMPATIBILITY

Throughout the history of both consoles, Sony and Microsoft have installed backward capability as a matter of routine, meaning that gamers could acquire the latest generation of consoles and still play past-generation titles. It looks as if this long-standing consumer-friendly philosophy is coming to an end. The Xbox One and the PS4 do not officially offer this feature. However, disappointed fans of the Xbox may benefit from a clever fix as Microsoft's Albert Penello was recently quoted online saying that backward compatibility might not be completely lost on the Xbox One as users of Microsoft's new Azure Cloud service could utilise its ability to mimic the processing power of the Xbox 360. Likewise, industry insiders recently hinted that it may also be possible to play older PlayStation titles using Sony's newly acquired Gaikai cloud streaming service, running it as an emulator. Frankly, this is all speculative at the moment, but if it either process become a reality, it will open up another world of historic gaming content to fans of the respective devices.

THE VIRTUAL GLOVES ARE OFF

Who will win in the battle between these two giants of the console gaming industry? Only time will tell, but one thing is for sure, we are now entering the period considered by the industry as the eighth generation of the gaming console. The PlayStation and Xbox bloodlines are in their fourth and third generations respectively and probably as closely matched as they have ever been in specs. The Sony does have a slight edge on sheer grunt, but both consoles have peripherals and features that enrich gaming and multi-media experience far beyond just bashing buttons. What will decide this ultimate gaming battle will be the developers, and the way they employ the abilities of these two powerful machines. The titles that appear on each console will determine what camp the next generation of gamers choose to join. Gaming brand loyalty goes far beyond just Sony and Microsoft – gaming franchises like *Final Fantasy, Forza, Halo, Call Of Duty, Killzone, Dead Rising, Need For Speed* and *Tekken* will be the decisive factor. Interestingly there are many titles that will appear on both units, but for existing console gamers it will be the brand specific titles that will have the most influence over which console they ultimately choose.

But there is one other thing to consider in all of this: what have the likes of Nintendo been developing while the eyes of the world have been turned on Sony and Microsoft? Another contender may yet enter the ring. Nintendo have done it before with the Wii, and although the Wii U has experienced relatively poor sales compared to past Nintendo products, this isn't down to the hardware or its specs and owes far more to the paucity of decent software to run on the console. If developers get behind Nintendo again, who knows what could happen? But until then it is seconds out for round one of this new, exhilarating battle of the console wars.

' The big development though is that it's been designed to utilize the processing power within the cloud for many of its games, like the racing heavyweight Forza V. '

The Gadget Show's
Gaming Top Ten

Tetris was Game Boy's first 'killer app'. It was included as a 'pack-in title' in every global region except Japan and was key in ensuring the revolutionary hand-held became the global 'must-have' gadget of the late 1980s. In the US, the initial shipment of one million consoles sold out within a matter of weeks, and more than 300,000 units sold on the first day in Japan.

Its simple and yet addictive game play plus repetitive but catchy theme tune was an instant success. Players manipulate falling shapes (Tetrinos) to get them to mesh together correctly and in doing so 'clear lines'; it still captivates gamers today and even becamethe basis *of one of The Gadget Show's* world record attempts!

Tetris and Game Boy

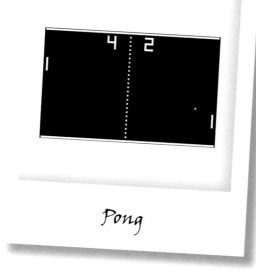

Pong

The Nolan Bushnell game that in 1972 kick-started both the home gaming market (with the Magnavox Odyssey console) and planted the seeds for early arcade cabinet domination. There were clones and even a few lawsuits along the way, but without this simple game of bouncing a sprite between players' paddles, the latest generation of graphic-intense sophisticated gaming we enjoy now would simply not exist.

Street Fighter II: The World Warrior

Universally acknowledged as the title that started the whole 'fighting' game genre. Released for the arcade in 1991, *Street Fighter II* was the first title to use 'special moves' and a six-button controller configuration. It offered multiple playable characters, each with their own style and moves.

By 1993, sales of *Street Fighter II* had exceeded US$1.5 billion and other software houses were busy developing their own fighting titles. It's still cited as Capcom's best-selling consumer game of all time, and is just as much fun to play today!

It put mobile gaming firmly on the map! *Angry Birds* is a video game franchise created by Rovio Entertainment. The first game was released for the Apple operating system (iOS) in December 2009 and since then in excess of twelve million copies of the game have been bought from Apple's App Store. Other smart phone-based versions followed, including Android, Symbian and Windows Phone incarnations.

The franchise has now expanded to include video game consoles and PCs. A global gaming and merchandising industry now worth in excess of US$35 billion dollars.

Angry Birds

The world's most played massively multiplayer online role-playing game (MMORPG) that transformed the online gaming community forever. An immense and immersive gaming experience, this was actually the fourth release in the series set in the fantasy Warcraft universe, and was released in 2004 on the tenth anniversary of the Warcraft franchise.

With over seven million subscribers in 2013, *World of Warcraft* is currently the world's most-subscribed MMORPG, even holding the Guinness World Record for the most popular MMORPG by subscriber numbers alone, and its might is still growing.

World of Warcraft

The two characters have come to represent the Sega and Nintendo brands respectively and are instantly recognizable today, even to non-gamers. Mario has featured in more than 220 games since his *Donkey Kong* debut in 1981 and the marketing appeal of the Italian plumber remains strong to this day.

Ten years later Sega's Sonic the Hedgehog emerged and became an instant hit. His series of titles has now sold more than eighty million globally, and with spin-offs such as a dedicated cartoon series his appeal as an iconic character reaches far beyond gaming. In 2005, Sonic, alongside his long-time rival Mario, were inaugural game character inducted into the San Francisco 'Walk of Game'.

Sonic and Mario

Nintendo Wii

One of the most innovative additions to the gaming market that concentrated on 'gameplay' rather than pure graphics; it meant anyone could play from the ages of eight to eighty and beyond. The built in motion sensors offered a whole new means of gaming control and the Wii immediately invented the invisible tennis racket, golf club, laser rifle and sword.

Each player could feel part of the Wii global family by creating their own personalised 'Mii' character. Wii changed the way we interact with games forever, and was also responsible for the occasional over exuberant destruction of overhead lights, ornaments (and if the wrist strap wasn't used) television screens!

This has to be the best known arcade and computer game of all time. Originally developed by Namco and first released in 1980, it became synonymous with 1980s popular culture. It was the first game that managed to extend its commercial reach with merchandise. In a market dominated by space-shooters, it had a mass appeal for both genders, becoming one of the highest-grossing video games of all time.

It is estimated that aside from its seven million plus cartridge sales on the Atari 2600 gaming console, and the 1.5 million Coleco's tabletop Mini-Arcade versions, the 350,000 original arcade cabinets absorbed more than US$2.5 billion in quarters alone by the mid 1990s!

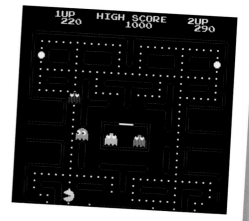

Pac Man

It was one of the first mainstream home computers and British too! The inventor, Clive Sinclair, was even given a knighthood for his entrepreneurism in bringing the colour enabled 8-bit personal computer into the homes of over five million users worldwide.

With limited memory and running Basic programming language, it was said to have been the catalyst for the UK IT and software gaming industries, and a staggering 24,000 software titles have been released for it since launch. Amazingly, new titles continue to be released with over one hundred in the last two years alone! It's a must have device for any retro gaming fan!

ZX Spectrum

It took the established first-person genre to a new level of realism and immersive gameplay. The title would sell millions of copies with each new release. The franchise has two titles in the top four bestsellers in the all- time Xbox 360 chart, and has globally sold in excess of one hundred million copies. With over forty million monthly active players across all of the *Call of Duty* titles, and over ten million *Call of Duty: 'Elite'* users, over 1.6 billion hours of online gameplay have been logged for *Modern Warfare 3* since its release in 2011.

Originally designed as a direct competitor to EA's *Medal of Honour,* instead of concentrating on a single character's plight, *Call of Duty* allows the player to see all aspects of the conflict by controlling American, British and Russian troops – each in their own dedicated story-driven missions. A game series that has become one of the most successful entertainment franchises in history.

Call of Duty Series

Build Your Own
Phone Charging Trousers

Jason and Rachel build a pair of trousers that can charge your mobile devices while you walk. A brilliant idea, but is it practical? The pair head for the streets of Birmingham, where they test their phone-charging trousers on the failing batteries of the general public.

List of Build Items
Stepper motor from printer
Solar boost charger
Metal rods plus gearing
Capacitors, diodes and circuit boards

All mobile devices need energy and that mostly comes from rechargeable batteries. Fine when you are at home or work and can plug your device into the mains, but what happens when you are out and about? Of course humans have lots of energy and in this ingenious build, Jason and Rachel come up with a way to harness the energy of day-to-day activity and use it as a source to charge those ever power-hungry gadgets.

A PRINTER MOTOR SITS AT THE HEART OF THE CHARGER.

This is an electric motor from a printer. When turned manually it can generate between 3 to 24 volts of alternating current (AC).

THE TWO METAL RODS WILL RUN DOWN RACHEL'S TROUSER LEG

We harnessed this energy generating potential by sewing the stepper motor into the knee joint of a pair of Rachel's pants. The motor was connected by a simple gearing set up to two metal rods, one running up and the other running down the inside leg of her trousers with the motor acting as the joint in the middle. As Rachel walked, the rods drove the motor via a simple gearing set up and thereby generating an alternating current with each step.

However, phones and other electronic devices don't use AC that constantly fluctuates between positive and negative voltages; they use direct current (DC), which is a constant positive voltage. To deliver this to the phone the AC current was converted using a simple circuit board created out of diodes and a 25v capacitor.

Once the AC current was converted to DC we then needed to smooth out the voltage to deliver a constant five volts, which is what USB powered devices require. This involved another link in the circuit called a voltage regulator to ensure a smooth five volts would be delivered to the charging phone.

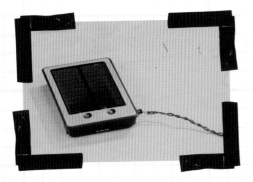

This alone would work to charge the phones. However, once Rachel stopped walking, the phone would no longer charge. To counter this a solar charger was added to the circuit. This also provided us with a USB socket for the phones to attach to. So when Rachel moved the electricity generated topped up the solar charger's battery which then in turn charged the phone. Simple!

And yes it worked!

Out on to the streets and to the amazement of the public, Rachel's dance moves soon had the energy-strapped citizens of Birmingham queuing up to charge their phones!

JASON EXPLAINS THE MECHANICS OF THE CHARGER TO RACHEL.

IT REALLY WORKS! A COMBINATION OF PHYSICS, MECHANICAL ENGINEERING AND SOME TAILORING AND YOUR PHONE WILL NEVER RUN OUT OF BATTERY POWER AGAIN.

Vintage tech ads

'Give her a Hoover and you give her the best.'

Published circa 1938

While this ad may fly in the face of modern sensibilities when it comes to sexism, in the 1930s it was very much in keeping with the style and mood of most domestic technology adverts. Hoover had recently unveiled the Model 150, the most modern machine of its time, a machine that included such innovations as an empty bag indicator; automatic height adjustment; a magnesium body; instant tool conversion; and a two-speed motor. The cleaner sold from 1936 to 1939 and was priced at US$79 for the complete package, the equivalent of £1,500 (or five high-end vacuum cleaners) today.

The Gadget Show Hall of Records!

Never content with simply testing technology for its intended use, *The Gadget Show* can take considerable pride in the fact that over the past seventeen series the program has actually put its practical testing to a greater good by achieving eleven prestigious **Guinness World Records**! How many other television programs can claim that kind of track record? And where else have presenters pushed the boundaries of what is possible with technology, sometimes even putting their very own wellbeing on the line in the pursuit of gadget-powered greatness? We salute all of their endeavours and applaud the tech that has made it possible – from the fastest Scalextric cars, to a power tool-powered dragster and the jet-engine equipped street luge. With gadget-wise ingenuity, the show takes everyday technology and does something extraordinary with it!

In Gadget-Historical-Order The Records Are:

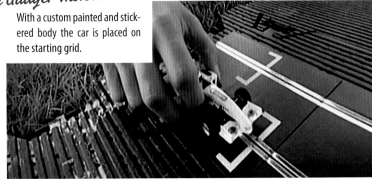

With a custom painted and stickered body the car is placed on the starting grid.

A jaw dropping 983.88 mph! Even if it was to scale.

1 It's Not The Size That Matters. The World's Fastest Slot Car

The show's first world record was achieved with a modified Honda F1 Scalextric car. After a few attempts the car reached an amazing speed of 983.88 scale mph (1,583.4 scale km/h) driven by Dallas Campbell at the Chatsworth Rally Show, Chatsworth on 6 June 2008. Ok, so it's not the biggest vehicle *The Gadget Show* has ever broken a record with, but it was the first – so has a warm place in our hearts!

Back in the studio, both parties get a reception fit for a proper F1 driver, and *The Gadget Show* has earned its first world record!

Head down while creating as little drag as possible, the car blasts past the camera with water shooting out twenty-metres behind it.

② Under Pressure. The World's Fastest Water Powered Car

Ignore solid fuel rockets, nitro methanol or jet engines because the show's second world record was powered by pure H20 and air pressure. With Jason Bradbury firmly strapped in, the unusually shaped vehicle hit a heady top speed of 16.65 mph (26.8 km/h) at Wattisham Airfield in Ipswich on the 15 March 2010. That may not seem so amazingly fast, but with nothing but displaced pressurised water powering the car to the record, it was quite an amazing achievement for *The Gadget Show* (and it was all thoroughly environment-friendly too).

Forget *Days Of Thunder* think more Jason and pressurised water.

The team behind the build congratulate an elated Jason, and celebrate acquiring world record number two for the show.

Hidden in the depths of the Bullring in Birmingham sixteen of the latest RC cars are lined up ready to try and break another Guinness World Record.

3 Flying Without Wings. The World's Longest Radio Controlled Car Jump

As many of you may already know, Jason is something of an RC nut. So when it was suggested that the show make an attempt on the RC car long jump record, his enthusiasm for this endeavour was only matched by the grin on his face! After trying (and failing) to achieve anything approaching the record distance with other vehicles, a brushless and LiPo powered 1/8th scale HPI Vorza Rally X car was lined up for a final attempt, and the rest is simply RC history. The longest ramp jump ever performed by a remote controlled model was recorded at 26.18 metres in Birmingham on 25 March 2010, and is just the first of two RC records the show has under its belt. FACT: Jason still has the car that secured the record and on occasions can be found pulling himself around his local park with it using a dog lead while standing on a skateboard!

When you see the expanse of Astro Turf laid out after the ramp, you get an idea of the distance the cars will have to travel for the record.

OK so that's not the Vorza Jon's holding, but you get the gist! Record number three is safely in the bag!

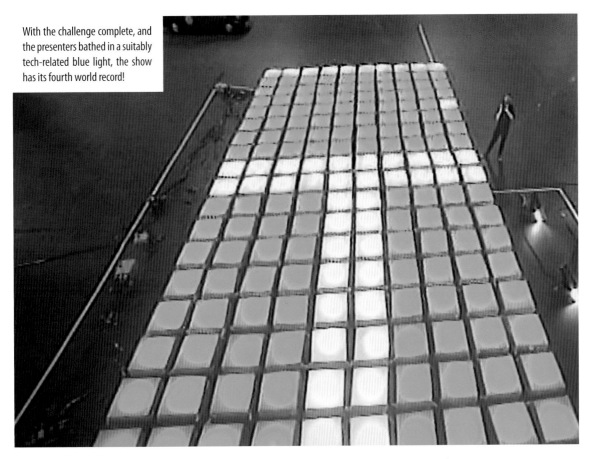

With the challenge complete, and the presenters bathed in a suitably tech-related blue light, the show has its fourth world record!

4 Go Big Or Go Home. The World's Largest Game Of Tetris

Forget the Nintendo Game Boy version, who would have thought that when Alexey Pajitnov created Tetris in 1985, twenty-five years later *The Gadget Show* would honour his programming skills by creating (and successfully operating) the world's largest Tetris computer game – measured at 105.79 square metres. It was so large that the whole presenter team was required to operate it, including one high up in a cherry-picker relaying direction to the other four below! This mammoth Gadget Show gaming record took place on the 15 September 2010.

Who would have thought that this amazing high-tech world record would start with over a hundred low-tech cardboard boxes?

The original 8-bit version of the games was used to power the pixel mapping software.

The presenters are shown into a warehouse and told they must work as a team to move the crane with just their minds.

5 'Come On Team, Concentrate!' The Heaviest Machine Ever Moved By Mind-Control Alone

Now the show's presenters aren't too short of brain-power, but they all put it to the test on 17 March 2011 when they successfully moved the heaviest machine ever using just a brain control interface. The presenters were required to work as a team to achieve it, with Suzy Perry giving *The Gadget Show*-hive-mind directions from a cherry picker, while the other four used mind-control headsets to each operate an axis or function to move a 56.2 tonnes (61.95 tons) crane and electro-magnet, to first pick up and then physically move a car from one location to another in a giant Studley warehouse. Mensa, be afraid, be very afraid . . .

Not only that, they must also pick up a car and move it to another part of the warehouse!

'We have lift-off!' Gadget one is lifted using nothing more than the power of *The Gadget Show*-hive-mind and one of the biggest cranes in the West Midlands!

The view as seen from a forward facing GoPro camera.

6 What A Rush. The World's Fastest Jet-Powered Street Luge

If it's got wheels and resembles any kind of board, Jason wants to have a go on it. If it's got wheels, resembles a board *and* has jet engines fitted, then Jason HAS to have a go! Having failed to achieve a world record on a previous occasion with an earlier incarnation of a jet powered street luge (during which Jason described himself as having 'a 110 mph Richard Hammond moment!'), in true Gadget Show style he revisited the record. On 9 August 2011, Jason successfully drove an upgraded version of the jet powered steet luge, reaching a staggering top speed of 115.83 mph (186.41 km/h) during the show's now legendary 200th episode. There was drama as at one point he hit full thrust and was for a while, when he said later, 'Just along for the ride!' But, oh boy, what a ride!

The original design concept for the jet luge that would, on its first attempt to set a record and while travelling at over 100mph, almost put Jason in hospital.

The Mk II jet luge was twin-engined and built mostly of carbon fibre. Far more stable at speed; a good thing really as it took real guts for Jason to get back on it and go for the record again!

After a failed challenge and technical issues with Polly at the wheel, Jon sets out with the help of Robo Challenge to fit the power tool-powered dragster with 'upgrades'.

7 How Many Chainsaws Does It Take To make A Dragster? The World's Fastest Powertool-Powered Vehicle

Now that may sound like the start of a joke, but when Grant Cooper and Robo Challenge get involved with a Gadget Show build, and there is a world record at stake, it doesn't get any more serious! This was another challenge re-visited; in an earlier attempt a power tool-powered dragster had suffered mechanical failure with Pollyanna in the driving seat at the annual Silverline Power Tool Drag Racing Championships. After a minor design tweak and now with six engines and far more power on tap, Jon Bentley was nominated as the pilot and strapped in behind the wheel. As part of the 200th episode and its trilogy of records, the dragster achieved a fastest run of 72.74 mph (117 km/h) at Santa Pod Raceway, Northamptonshire on 22 August 2011 – catapulting it, but fortunately not Jon, into the record books and gaining the show its seventh world record in the process!

It's *The Gadget Show's* 200th episode and Jon Bentley puts everything on the line for its reputation as the tech show that goes to the places no others dare!

Past the line and looking good! But had we done it?

Usually found in large scale RC jets singularly or in pairs, the combined thrust these six generate was simply astonishing!

Now that's some serious voltage! The LiPo packs used in this challenge are more commonly found in performance RC helicopters and planes. With six high thrust EDFs fitted, the voltage was definitely going to be needed.

The 'Black Box' but for nothing more sinister than controlling the maximum amount of acceleration and throttle available to Ortis during the run.

At 72mph the bike was rock solid stable and arrow straight. A good thing really!

8 Who Needs Pedals? The World's Fastest EDF Powered Bicycle

The last of the three records achieved during the 200th episode involved fitting thrust-producing tech to a vehicle again, but this time to a bicycle! With Ortis Deley suitably equipped with helmet and motorcycle leathers, and after overcoming some lower speed instability issues, Ortis dug deep and eventually took the Electric Ducted Fan (EDF)equipped bike to 72 mph (115.87 km/h) down the drag strip at Santa Pod Raceway on 24 August 2011. A leisurely Sunday ride this was not!

And the verdict? Yeah, that's right, Ortis has done it! Gadget Show world record number eight goes to an EDF jet-powered bike, amazing!

The world's biggest loop-the-loop ever attempted with an RC vehicle.

The best of the current breed of RC bikes are lined up for the challenge.

9 Jason's Goes Round The Bend. The World's Largest Radio Car Loop-The-Loop

The show's second RC world record started as an idea for *Gadget Show Live*. After it was shelved for logistical reasons, the show had another think about it and instead of using an RC car (as originally planned) the show's RC tech-guru Pete Gray supplied an E-Gyro equipped ¼ scale RC Moto X Bike, the VMX 450. With a loop measuring well over nine feet tall, on 15 June 2013 Jason Bradbury successfully piloted the bike around the loop that was situated in an empty warehouse in Birmingham.

It's unanimous: the VMX450 'Deegan' E-Gyro is chosen to attempt the world record.

Initially it was thought the record had been set at the first attempt, but on closer inspection of the video by the Guinness judges, it was noted that the bike had clipped a GoPro camera placed on the down ramp. After re-grouping, Jason made three more attempts, and finally took the bike cleanly around the loop, thus securing the show's ninth world record.

Jason cross-checks all the measurements to make sure we are legal and correct.

So, what next? Where can the show take its thirst for taking every day technology and achieve the impossible with it? With the always-enthusiastic team of tech-head researchers, producers and presenters at *The Gadget Show* anything is possible. Just watch this space, as something with a jet engine, or a power tool strapped to it may just fly, drive or hover past any second now . . .

After a few tense moments and several failed attempts: 'Yes! We've done it' says Jason.

Getting both Polly and the JetLev ready for the record attempt. Love that smile!

The start of the record attempt in the idyllic Msida Yacht Marina in Malta.

Maintaining a good pace and altitude using just the power of water.

The route. GPS mapped and verified at 22.64 miles in total.

10 'Perfect, Now Just Stay Airborne For 20 More Miles'. JetLev Distance Record

For the 250th episode on 8 November 2013, we decided to establish a record for the longest journey on a water-powered jet pack! Clearly we needed a global thrill-seeking presenter to take on this challenging task. Step forward Pollyanna Woodward, our presenter with an international 'license to thrill'. The show travelled to Malta where Polly would start her journey on the neighbouring island of Gozo, loop around the small island of Comino, before finishing at the country's capital, Valletta. The machine she would be using was the JetLev Flyer, something Polly has tested in 2011.

The JetLev is a cleverly designed machine: a floating pump sucks up seawater, forcing it up a ten metre hose to a pair of high-pressure nozzles on the back of the operator. Arms at the front of the pack allow the direction of the nozzles to be manipulated, pushing the pilot upwards, forwards or sideways. There is a twist grip throttle to adjust the amount of thrust. To make sure the record attempt was legitimate, a GPS logger had been fitted inside the JetLev pump to record the exact course and distance – 36.45km or 22.64 miles.

What an achievement! Polly, overcome with emotion, jets into the record books!

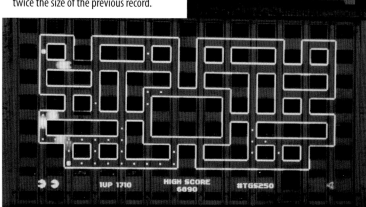

The final projected game at night, almost twice the size of the previous record.

The Millennium Mills in daylight.

Shay Moradi from Running In Halls coded the game for us.

An oversized arcade game needs and over-sized arcade stick!

And yet another world record is in the bag.

11 Go Big Or Go Home. Large Scale Projection Gaming Record

The records for the largest architectural projection-mapped game now stands at 23,881 square feet (2,218.65 square metres) achieved by *The Gadget Show*, at the ExCel Centre, London, UK on 18 November 2013.

To make the show's 250th episode even more memorable, we decided to do something visually stunning. Rachel, and Jason Bradbury were chosen for this particular challenge along with the 1980s classic, Pacman. The projection would faithfully recreate the original graphics and sounds of Pacman, and be beamed across the Thames on to one of the largest mill complexes ever to be built in London – the Millennium Mills.

Using the natural architecture of the building itself, the position of the windows and the building's sheer scale, programmers coded a Pacman maze – the size of which had never been seen before. A huge game requires an equally huge controller, so we armed Jason and Rachel with a suitably-sized arcade joystick created especially for the event. Once all the elements were in place, it not only looked amazing, but proved to be just as addictive as the original arcade game. Our biggest tech challenge so far paid off, and the series marked its 250th episode in true Gadget Show style by receiving its eleventh Guinness World record.

Vintage tech ads

'Undo. Windows.'

Published circa 1983

Initially launched just for Xenix and IBM PCs, and later for Windows-based machines in 1989, Microsoft Word is still the world's favourite word processing software. Free demonstration copies of the DOS application were bundled with the November 1983 issue of *PC World*, making it the first disk distributed on a magazine. Fast forward to today and the Microsoft Office suite, that Word is part of, has more than 750 million global users, ensuring the phrase, 'I'll send it to you as a Word document.' is commonplace.

The Gadget Show's
Transport Top Ten

The Ford Model T has been described as the world's most influential car of the twentieth century and it is amazing how little the car itself and production methods have changed since the first example rolled out of the factory on 27 September 1908. Henry Ford's visionary assembly line production that replaced hand crafting, along with highly efficient fabrication techniques for the time, meant that just nineteen years later, he and the world watched the fifteen millionth Model T Ford drive away from the Ford factory in Highland Park, Michigan.

Henry Ford remarked: *'I will build a car for the great multitude. It will be large enough for the family, but small enough for the individual to run and care for. It will be constructed of the best materials, by the best men to be hired, after the simplest designs that modern engineering can devise. But it will be so low in price that no man making a good salary will be unable to own one – and enjoy with his family the blessing of hours of pleasure in God's great open spaces.'*

Model T Ford

Macmillan Velocipede

The idea of riding on a two-wheeler actually pre-dates Kirkpatrick Macmillan's 1839 invention by many years, but what Macmillan, a blacksmith by trade, did was to add simple mechanics that meant the rider could propel the velocipede with pedals instead of simply pushing the machine along with feet on the ground. Macmillan's design wasn't overly complicated. Using his ingenuity and blacksmithing skills, he constructed the velocipede mostly of wood with iron-shod wheels, a fully steerable wheel at the front and a larger wheel at the rear driven by pedals linked by connecting rods.

It was responsible, allegedly, for one of the earliest recorded vehicle offences when in 1842, a Glasgow-based newspaper reported an accident in which an anonymous 'gentleman from Dumfriesshire, bestride a velocipede of ingenious design', knocked over a pedestrian and was fined five shillings. A plaque on the Macmillan family smithy reads: *'He builded better than he knew . . . '*

Bell Jet Pack

The hydrogen peroxide powered Bell Jetpack featured in the James Bond film *Thunderball* and during the US Olympic opening ceremony in 1984 capturing the imagination of millions. Who doesn't dream of unencumbered flight? After many years of development and numerous 'tethered flights', on 20 April 1961 (the week after Yuri Gagarin's famous space flight) the first free flight in the history of a jet pack took place on an open space by Niagara Falls airport.

Harold Graham reached an altitude of approximately 1.2 metres and flew forward at a speed of 10 km/h. Graham flew 108 feet (less than 35 metres) before landing safely. The entire flight lasted just 13 seconds, but was a moment when sci-fi became reality. During the years since the original Bell pack flew, Bell scientists and in-house engineers have made many improvements in the power-to-weight ratio of the unit, utilising new composite materials as they have become available and enabling pilots to carry unaided a pack with more propellant.

The latest incarnations will fly around fifty percent longer than the original 1960s units. That's sounds pretty impressive until you realise that it means you get to fly for all of thirty seconds instead of the fifteen seconds or so back then!

Segway

The Segway is a ground-breaking piece of tech and an innovative, modern, environmentally sensible take on transportation. But the Segway has, and will, continue to divide opinion. It can make users look (and feel) somewhat ridiculous. Regardless, it cannot be denied that the Segway is a unique transportation device. Renowned inventor Dean Kamen founded his company with a vision to develop highly efficient, zero-emission transportation solutions using 'dynamic stabilization' technology. The company's research and development focused on creating devices that required minimal space, were extremely manoeuvrable and could be operated on pedestrian sidewalks and pathways.

The Segway Personal Transporter (PT), the world's first electric, two-wheeled, self-balancing transportation device, was unveiled in September 2001. By 2002 the first State bill in the US allowing Electric Personal Assistive Mobility Devices (EPAMDs) to operate on sidewalks was signed into law in the company's home State of New Hampshire. Within twelve months a total of thirty-one States had passed laws allowing the Segway PT to operate on their sidewalks. Segways are currently banned on British roads, but are legal on private land.

The Segway can deliver an impressive energy efficiency equivalent to 450mpg. The Patrol models with security personnel or police on board are often seen at airports and in the centre of major cities. The original Segway has evolved and current models include the i2 Commuter and the x2 Adventure. Tragically, the owner of the Segway's manufacturing facility died in a fall from a cliff while out riding one of his vehicles in 2010 .

Specialized Stumpjumper

SUMPAC

Built in 1981, the original Stumpjumper was the first proper mountain bike that could be bought from a bike shop. Pre-Stumpjumper, the team behind the design were riding dirt roads and smooth trails on their road bikes. Specialized's founder Mike Sinyard said that the company's aim was to build a mass-produced machine that rode and looked as though it was a custom built bike.

The first Stumpjumpers had welded steel frames, a modified BMX stem and handlebars based on those seen at the time on off-road motorcycles. Equipped with 15-speed Suntour ARX GT gears, the design also featured cantilever brakes and a TA Cyclotourist chainset, all components originally designed for touring bikes. The bike weighed in at around 14kg and sold for US$750 as a complete bike or US$395 for the frame only. Specialized marketed it as an affordable and versatile bike for the new emerging sport of mountain biking.

Initially retailers were sceptical; however the first shipment sold out in just six days and its introduction is cited as one of the main contributing factors in the rapid rise in popularity of mountain biking. An original Stumpjumper is displayed in the Smithsonian Institute in Washington, D.C. In 2007, to mark the bike's 25th anniversary, Specialized produced a limited edition version of the original, but with a twist (as it featured modern components), calling it the Stumpjumper Classic.

Built by a group of post-graduate students in Southampton and funded by the Royal Aeronautical Society, in a single flight it achieved what man has been striving towards for a thousand years. The Southampton University Man Powered Aircraft (or SUMPAC) took to the skies on 9 November 1961, and became the first human-powered aircraft to make an officially authenticated take-off and flight. Originally designed and built for an attempt at the much sought after Kremer Prize, it sadly failed to complete the one-mile official 'figure-of-eight' course, thus missing out on the £50,000 prize money!

The aircraft constructed from balsa, plywood and aluminium, was originally covered with doped nylon in an approach familiar to all schoolboy aircraft modellers. It was powered using pedals and chains to drive a large two-bladed pusher propeller. Piloted by gliding instructor and test pilot Derek Piggott, its first flight was at Lasham airfield and covered a distance of 64 metres while climbing to a height of 1.8 metres.

A total of forty flights were made by SUMPAC, but after crashing in 1963 it was decided to retire the aircraft.

McLaren F1

It all started in an airport lounge in Milan in 1988. While waiting for a delayed flight, McLaren F1 team boss Ron Dennis, business partner Mansour Ojjeh, McLaren technical director Gordon Murray and head of marketing Creighton Brown were discussing how best to harness the company's immense technical capability. Despite almost total domination of the season's Formula 1 World Championship (winning fifteen out of sixteen races), it was clear to Dennis that a sustainable business could not rely on motor racing alone. Among a range of suggestions the idea of building the best sports car in the world emerged. In 1989 a new company, McLaren Cars, was established and in March 1990 an infamous ten-hour design meeting resulted in the commitment to build the fastest and best-handling supercar in the world.

It would have the highest power-to-weight ratio of any production car and be sufficiently practical for everyday use. The first XP model was unveiled to the world's press in Monaco before the Grand Prix in May 1992. Levels of interest were huge and the bold claims made for its performance suggested a new type of supercar had arrived. At 1,140 kilograms, the F1 was much lighter than any of its competitors and the engine produced more power too.

It could accelerate from 0-60 in 3.2 seconds, 0-100 in 6.3 seconds, 0-150 in 12.8 seconds and 0-200 in 28 seconds and was so flexible that it could accelerate cleanly, without jerking or shuddering, from 30mph to 225mph in sixth gear. McLaren had achieved its objective to produce the ultimate supercar. There was a price to pay for this no-compromise approach to engineering that included a gold foil-lined engine bay (each F1 has about twenty metres of gold foil for the best possible thermal insulation). The £634,500 price tag was as astonishing as the car itself, but that didn't stop the orders from rolling in!

North American X-15

The X-15 was a rocket-powered aircraft operated by the United States Air Force (USAF) and the National Aeronautics and Space Administration (NASA) as part of the legendary X-plane series of experimental aircraft. Piloted by test pilots and soon-to-be astronauts like Neil Armstrong, the X-15 set speed and altitude records in the early 1960s, reaching the edge of space and returning with valuable data that would be used in aircraft and spacecraft design. Armstrong made his first X-15 flight on 30 November 1960 in the number one X-15 and his second flight on 9 December 1960 in the same aircraft.

This was the first X-15 flight to use the ball nose, which provided accurate measurement of air speed and flow angle at supersonic and hypersonic speeds. The X-15 employed non-standard landing gear. It had nose gear with a wheel and tyre, but the main landing gear consisted of skids mounted at the rear of the aircraft that meant it could only land on a dry lake bed rather than the usual concrete runway. The X-15 still holds the official world record for the fastest speed ever reached by a manned aircraft. Its maximum speed was a staggering 4,510 miles per hour (7,258 km/h) and was piloted by William J. Knight on 3 October 1967.

When you consider that the second fastest aircraft of all time is the Lockheed SR-71 Blackbird that flew at half that figure you can see just how awesome the X-15 was.

Satellite Navigation

Automotive Satellite Navigation systems (sat nav to you and me) typically use a Global Positioning System (GPS) to acquire positional data – longitude, latitude and even height – and then combine it with the unit's map database. As of April 2013 only the United States NAVSTAR GPS and the Russian GLONASS are actually globally operational. China is in the process of expanding its regional navigation system into the global Compass navigation system and the European Union's Galileo positioning system is in initial deployment phase, scheduled to be fully operational by 2020. Other countries like France, India and Japan are also developing regional navigation systems.

Global coverage for each system is generally achieved by a satellite constellation of between twenty and thirty medium Earth orbit (MEO) satellites, spread between several orbital planes. The actual systems vary, but all use similar orbital inclinations and orbital periods of roughly twelve hours. Most sit above the Earth at least 20,000 kilometres (12,000 miles) away. With the advent of GPS-enabled smart phones and sat nav devices fitted to many cars there lurks a danger.

The devices and the routes they suggest are only as good as the stored data, and our increased reliance on sat navs has been blamed for causing more than a few incidents and accidents. Over-reliant drivers have driven straight into canals and rivers or, infamously, attempted to drive across the muddy Moreton Bay Australia at low tide on a non-existent road heading for a small island.

Cats Eyes

There is a good deal of urban myth surrounding the invention of the reflective road stud. Some say inventor Percy Shaw's inspiration came from seeing the reflection of cats' eyes in his car headlamps, but during a 1968 interview with Alan Whicker, Shaw told a different story of an incident in 1933. On a dark, foggy night he was driving down a steep winding road from Queensbury to his home in Boothtown, Yorkshire – a journey he had made many times before. Previously, he'd followed the reflections of his own car headlights on the highly polished tramlines set into the road (there were no streetlights in those days).

The reflections were a great help while negotiating the hazardous sweeping bends. On this occasion he was suddenly and without warning plunged into pitch darkness as the reassuring reflective light disappeared. A whole section of the tramlines he had hitherto relied upon had been taken up for repair. Shaw realised the potential of using a similar system to improve road safety more widely. He would need to create a reflecting device that could be physically fitted into the road's surfaces and accurately show its direction of travel. After many trials Shaw took out patents on his invention in April 1934 and in March 1935 Reflecting Roadstuds Ltd was incorporated, with Percy Shaw as managing director.

The company would go on to produce more than a million roadstuds per year, exporting them globally. In 1965 he was awarded an OBE for his services to exports, and in 2005 was listed as one of the greatest Yorkshire people to have ever lived!

ULTIMATE GAMING GADGET FACE-OFFS

Atari was formed in 1972 by Nolan Bushnell and Ted Dabney who can be credited with creating the video gaming industry through the launch of the simple but addictive game 'Pong'. Six years later, Taito's Space Invaders became the world's first global blockbuster – and the rest is gaming history.

Even though the popularity of arcades continued well into the 1990s (and remains hugely popular in Asia to this day), it was during the 1980s that gaming moved out of the arcade and into the home. The arrival of the first generation of gaming consoles along with personal computers meant that the powerhouse that is the gaming industry had well and truly arrived.

As we enter a new age with the latest generation of consoles, the global sales of major gaming titles in the week of release will often beat the opening weekend of even the biggest Hollywood blockbusters.

In this section we take gaming heavyweights and let them face-off in a fight that isn't just about having the latest hardware, or the best graphics. It's about the playability of each device and the power of the iconic gaming characters and franchises that we have all grown up loving.

3DS

Success: 2
Nintendo announced a significant price reduction from US$249 to US$169 amid disappointing sales.

Lifetime: 3
2011 to now, but could well be replaced by smart phone games.

Innovations: 4
The first mainstream handheld gaming device to use 3D without glasses.

Fun Facts
Nintendo started experimenting with 3D technology as far back as the 1980s with the Famicom 3D System.

Wii

Success: 5
50,000 consoles were shifted by lunchtime on the first day of its release.

Lifetime: 3
2006 to 2012.

Innovations: 5
Allowed you to control the game through real-life gestures and bodily movements.

Fun Facts
The two letters 'ii' in Wii are meant to resemble two people standing side-by-side or players gathered together and also represents the Wii Remote and Nunchuk.

GAME BOY

Success: 5
Game Boy and Game Boy Color combined have sold more than 110 million units worldwide.

Lifetime: 4
1989 to 2005.

Innovations: 4
The first truly successful, popular and well-known handheld console.

Fun Facts
During a routine train ride in the late 1970s, Nintendo games designer Gunpei Yokoi spotted a fellow passenger punching the keys of his electronic calculator in an effort to distract himself during a tedious journey. The experience resulted in Yokoi designing a Game & Watch range of portable games that ran from 1980 until 1991. It would eventually lead to the development of the Game Boy.

ATARI

Success: 4
Initial sales were poor but by Christmas 1979 Atari was selling millions of the device.

Lifetime: 4
Launched in 1977. Retired in 1992. The longest living console in gaming history.

Innovations: 5
The godfather of all gaming consoles.

Fun Facts
When Space Invaders was released for Atari, sales of the console doubled overnight.

SONIC THE HEDGEHOG

Success: 4
In 1998 Sonic the Hedgehog was voted the UK's most popular video game character.

Lifetime: 2
Not had a hit for quite some time now.

Innovations: 4
The Sonic Spin move. As well as the more traditional jumping and ducking gaming moves, Sonic could spin like a ripsaw blade to attack enemies.

Fun Facts
Sonic the Hedgehog was coloured blue to match the logo of the character's games publisher Sega. And his shoes were based on Michael Jackson's boots apparently!

CALL OF DUTY

Success: 5
Call of Duty: Black Ops 2 generated US$1 billion in sales in just fifteen days on release.

Lifetime: 4
2003 to now.

Innovations: 4
Role-playing game levels within multiplayer.

Fun Facts
Activision Blizzard created the Call of Duty Endowment (CODE), a non-profit foundation. It helps find employment for U.S. military veterans. 3,000 copies of Call Of Duty: Modern Warfare 2 were presented to the US Navy by CODE on 30 March 2010. They were delivered to over 300 submarines and ships as well as to Navy Morale, Welfare and Recreation facilities.

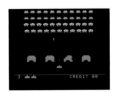

SPACE INVADER

Success: 5
It has generated more than US$500 million in sales on various different platforms and was one of the forerunners of modern video gaming. It helped expand the video game industry from a novelty to the global industry it is today.

Lifetime: 4
Space Invaders, as a franchise, has lasted for more than twenty years.

Innovations: 4
One of the first arcade games, it has been the inspiration for countless other video games. The 1980 Atari 2600 version ultimately quadrupled the system's sales and became the first 'killer app' for video game consoles.

Fun Facts
The game was so amazingly popular in Japan that it caused a coin shortage until the country's Yen supply was quadrupled.

MARIO

Success: 4
During the 1980s research showed that more school children recognised Mario than Mickey Mouse.

Lifetime: 5
Mario has appeared in over two hundred games and has been adopted as the publisher Nintendo's logo.

Innovations: 4
Mario has successfully made the move from 2D to 3D.

Fun Facts
Mario was originally called Jump Man, but was renamed after a landlord named Mario burst into a game designers' brainstorming session brandishing his cigar and demanding the rent!

HALO

Success: 5
Halo 4 topped sales of US$220 million worldwide in the first twenty-fours hours following its release.

Lifetime: 5
2001 to now.

Innovations: 4
Concentrated on system link and live multiplayer modes.

Fun Facts
At the Xbox Reveal on 21 May 2013, 343 Industries announced that a live-action television series of Halo would be produced as part of Xbox One's original TV programming, with Steven Spielberg serving as executive producer.

ANGRY BIRDS

Success: 5
There have been more than 1.7 billion downloads across all platforms since its launch.

Lifetime: 3
2009 to now. Four years and still going strong and soon the franchise will extend to an animation series and movie.

Innovations: 5
One of the first big successes of the new markets for tablet and smart phone apps. Simple, addictive and cheap, it redefined how successful games are made and published.

Fun Facts
The Angry Birds Theme Park is to open in May 2014 at Johor Baharu City Centre in Malaysia.

WHEN SCIENCE FICTION BECOMES SCIENCE FACT

> 'ONE COULD WRITE A HISTORY OF SCIENCE IN REVERSE BY ASSEMBLING THE SOLEMN PRONOUNCEMENTS OF HIGHEST AUTHORITY ABOUT WHAT COULD NOT BE DONE AND COULD NEVER HAPPEN.'
>
> **ROBERT A. HEINLEIN**

In 1902, Marie-Georges-Jean Méliès, a French illusionist and filmmaker, directed and starred in the first-ever science fiction movie, *Le Voyage dans la Lune (A Trip to the Moon)*. It was an epic that lasted all of seventeen minutes, and enthralled audiences at the time with its out-of-this-world revolutionary cinematography and the now famous 'rocket landing in the eye of the man in the moon' sequence. Since then science fiction literature and movies have been embroiled in a mutual and long-lasting love affair – sometimes with impressive far-seeing visionary results, but often with considerably less panache and a good deal of wrong-headed science along with unfeasible tech to boot.

Science fiction has always been written and read with its predictive quality sitting (often uneasily) at the core of the narrative: dystopian, utopian, operatic, exploratory or as a conducted tour through alternative histories of humankind; the writer is liberated from the constraints of the quotidian world we live in, if not post-Einsteinian physics – although that can quite often go by the board as well.

How well an alien world or strange new universe is conceived and realized on screen is a perfectly good way to judge the success of a sci-fi movie, as well as its power to captivate and entertain; but asking how well it presents a 'one day in the future' scenario, and whether the underlying science even remotely stacks up is also something that exercises the minds of tech-savvy audiences and critics.

Travelling faster than the speed of light, putative wormholes notwithstanding, is simply disallowed by every tenet of actual and perfectly well-understood laws of physics. There is enough genuine and exciting weirdness that exists either in the very biggest structures, or at the unimaginably small scale where foamy quantum space-time operates, in our mind-boggling 13.7 billion year old universe. Writers and film makers do not need to subvert things with tech and gadgets that appear to operate purely by pseudo-science (or magic).

So let us have a look at some of the most successful ever sci-fi movies, and discover to what extent the envisioned tech has become a reality.

Shaun Sparkes Design

MINORITY REPORT 2002

Directed by Steven Spielberg and based on a short story by master sci-fi writer Philip K. Dick, *Minority Report* is set in a future where a special police unit is able to arrest murderers before they commit their crimes.

THEME Crime Prediction

WHERE Memphis, USA

MOVIE TECH Pre-Crime

REAL LIFE TECH IBM CRUSH

'Criminal Reduction Utilizing Statistical History' is an IBM predictive analytics system that attempts to predict the location of future crimes. This state-of-the-art crime computer system was developed as part of the Blue CRUSH program in conjunction with Memphis Police Department and the University of Memphis Criminology and Research department. It monitors data, which its analytical software then extrapolates from and tries to predict trends, allocate resources and help to identify 'hot spots', thus aiding in a reduction in crime rates. It directs patrols, traffic enforcement, task forces, operations, high-visibility patrols and targeted investigations. Thanks to this software serious crime has reduced by more than thirty percent, including a fifteen percent reduction in violent crimes since 2006. It may not be able to actually see into the future like Agatha and the other 'Precogs' in *Minority Report*, but it does do a great job of predicting it!

THEME Gesture Controlled Computing

WHERE California, USA

MOVIE TECH Anderton's Computer Interface

REAL LIFE TECH Oblong G-Speak

A unique system that allows interactaction with 3D imagery and manipulation of computer interfaces as seen in *Minority Report*. Physical hands movements

offering spatial, networked, multi-user, multi-screen, multi-device computing environments!

THEME Flying People
WHERE Christchurch, NZ
MOVIE TECH Pre-Crime Jetpacks
REAL LIFE TECH Martin Jetpack

The Martin Jetpack consists of a purpose-built petrol engine driving twin ducted fans that produce sufficient thrust to lift the unit and its pilot and allow vertical take-off and landing (VTOL) plus sustained flight. Unlike the relatively short flight times of units like the Bell Jetpacks (seen in a James Bond film and at the US Olympics opening ceremony), the Martin unit can fly for 30 minutes and achieve speeds of up to 100km/h. The Martin Jetpack is the result of the inspired work of Glenn Martin and an enthusiastic team of engineers and experts, who have devoted many years to its development. It uses sophisticated composites and a highly efficient propulsion system to achieve the goal of personal flight, and comes with many safety features including a ballistic parachute. Originally designed with the leisure market in mind, the Martin Jetpack has found strong demand from a wide range of markets, including military, civil

are tracked by infrared cameras; zooming, panning, expanding and pulling are all actually possible. Oblong was founded in 2006 with the goal of creating the next generation of user interfaces, making computers more flexible, capable, useful, interactive, and empowering. Oblong's John Underkoffler actually designed the computer interfaces for the film and acted as its head science advisor. Today they sell commercial versions of the Minority Report computers and interface,

defence and recreation. The unit is in its final stages of development for commercial use in applications such as search and rescue, and as seen in *Minority Report*, law enforcement.

THEME Autonomous Transportation

WHERE Stamford, USA

MOVIE TECH Self Driving Car

REAL LIFE TECH Autonomous Google Car

Google claim that this tech will be on sale in less than a decade. The company has developed self-driving Prius and Lexus cars that have proved themselves to be far safer than human-driven vehicles. Data from Google's cars on public roads in California and Nevada showed that when a human was behind the wheel, the Google cars accelerated and braked significantly more sharply than they did when piloting themselves. The car's software also proved to be far better at maintaining the minimum safe distance between vehicles than human drivers ever could or would! Interestingly, Google aren't the only ones produce self-driving cars. Audi recently produced an example that competed in the gruelling Pikes Peak International Hill Climb, and Toyota, VW and GM are also in the process of developing them. Volvo has even built one that parks itself!

BACK TO THE FUTURE TRILOGY 1985 , 1989 AND 1990

Directed by Robert Zemeckis (*Forrest Gump* and *Cast Away*) and starring Michael J. Fox and Christopher Lloyd, the trilogy of films starts with a teenage Marty McFly (Michael J. Fox) accidentally sent thirty years into the past in a time-travelling DeLorean car (undoubtedly the other star of the movie) invented by his friend, Dr. Emmett Brown. Marty must make sure his own high school-age parents get together in order to save his own existence. Iconic tech and purposeful time travel ensures the films are deserving of the appellation 'classics'.

THEME Flying Car

WHERE Kelowna, BC, Canada

MOVIE TECH Flying DeLorean Time machine

REAL LIFE TECH Flying Car

The Maverick LSA 'Flying Car' is the result of six long years of research and development by a creative non-profit organization known as the Indigenous People's Technology and Education Centre. The Maverick LSA design has been developed as an easy-to-operate, air and land craft. It is intuitive and safe to fly, drive and

Shaun Sparkes Design

THE HITCHHIKER'S GUIDE TO THE GALAXY 2005

Conceived and written by the late Douglas Adams, *The Hitchhiker's Guide to the Galaxy* was broadcast originally by the BBC in the 1970s as a radio comedy. It would go on to become a beloved series of books and eventually a Hollywood movie starring soon-to-be Hobbit Martin Freeman as Arthur Dent who is, seconds before the Earth is about to be demolished by an alien construction crew, swept off the planet by his friend Ford Prefect, a researcher writing a new edition of the eponymous *Hitchhiker's Guide to the Galaxy*.

THEME	Real-time Translator
WHERE	USA
MOVIE TECH	Babel fish Universal Translator
REAL LIFE TECH	Real-Time Phone Translation

maintain by people in frontier areas, enabling them to use this unique vehicle in missions and humanitarian applications all over the world 'beyond roads.' These unique machines allow isolated and disadvantaged people living beyond the reach of roads, health care services, economic opportunities and formal education the opportunity to interact with fellow citizens of the planet. All without the use of a single flux capacitor, aluminium supercar or recycled garbage as fuel. Marty McFly and the Doc would be very impressed!

There is no doubt that computer-based translators are getting better – Google Translate is great – but AT&T has developed a real-time, person-to-person translation app for your smart phone! This incredible app helps you communicate across language barriers and currently operates in English, Spanish, Japanese, Chinese, French, German and Italian. It automatically recognises which language is being spoken and translates it in on the fly to a selected alternative language. So you'd like to have a conversation with a

THE MATRIX 1999

Keanu Reeves plays computer hacker Thomas A. Anderson, known in his hacking guise as Neo. Contacted by Morpheus (Laurence Fishburne), a legendary computer hacker branded a terrorist by the government, Neo is made aware that the real world is a ravaged wasteland where most of humanity has been captured by a race of machines that imprison their minds inside an artificial reality known as the Matrix. Neo must journey to the Matrix and confront the agents – super-powerful computer programs devoted to destroying Neo and a human rebellion.

THEME Virtual Reality

WHERE Japan

MOVIE TECH Simulated Reality

REAL LIFE TECH 8K TV

A lot of us now have access to a high definition television in our homes, and many will have seen a demonstration of the soon-to-be-implemented

friend who speaks Spanish, just speak into your smart phone and the app will first translate and then 'speak' your words in Spanish. It then listens to your friend and repeats the process in reverse. It's like having a personal interpreter on your smart phone. And the massive army of server-power behind the app means that it actually works! That has to be better than a Babel fish stuck in your ear as the Vogon Constructor Fleet destroys Earth to make way for a new hyperspace bypass . . .

broadcast standard 4K TV – and, no doubt, been suitably impressed. But right now, 8K is undergoing testing by a number of tech companies and broadcasters. To be clear, 8K has sixteen times the pixels of full-HD. Japan wants to start broadcasting in this format by 2020, and adding to it the extraordinary prospect of a surround sound experience that has no less than twenty-two speakers, and two subwoofers! Manufacturer Sharp recently demonstrated (at CES) an 8K LED TV with a screen measuring 85 inches. The TV was still a prototype, but its image quality was spectacular. With a resolution of 7,680 x 4,320, the demonstration videos showing children jumping off a bridge into a river and a train steaming out of a platform looked, well quite

frankly, real as real can be. By way of a comparison, 8K is roughly the equivalent of a 32-megapixel photo, a photo that usually contains enough data (pixels) to be blown up to billboard size. During the London Olympics, Japanese broadcaster NHK showed off the technology to visitors to the Olympic Park; audiences universally reported that the 8K experience was miraculously close to actually watching an event in the stadium. The firm has developed three cameras that can capture what the company is calling: 'Super Hi-Vision'. Currently filming at 60 frames per second (fps), the company aims to double that to 120 fps very soon. By contrast, here in the UK we currently broadcast HD TV programmes at a mere 25 fps.

2001: A SPACE ODYSSEY 1968

Perhaps one of the greatest science fiction movies of all time, Stanley Kubrick's film, with a screenplay by Arthur C. Clarke, is an exploration of the evolution of our species. Deep in the past, someone or something nudged evolution by placing a monolith on Earth and presumably elsewhere in the universe. Humankind eventually reaches the moon's surface, where yet another monolith is found – one that signals to the monolith makers that humankind has evolved enough for space flight. Now a journey to Jupiter is made to locate the alien intelligence that built the monoliths.

THEME Travel Technology

WHERE New Mexico

MOVIE TECH Space Tourism

REAL LIFE TECH Virgin Galactic Spaceship Two

Forget travel agents, try using a real life 'Accredited Space Agent' instead. Holidays are no longer limited to a week in Spain or a transatlantic jaunt to Florida. Within the next few years you could actually be heading into space. Well that's if you've got US$250,000 to spare for the ticket! Since 2005 Virgin Galactic has been asking aspiring astronauts to pay a deposit to reserve their place on 'Spaceship Two', the company's re-useable spaceship. If you do take the plunge then you will be joining a community of over six hundred would-be future-astronauts, and joining what is fast becoming one of the most exclusive travel clubs in the world. Prior to launch each passenger will have to endure a three-day pre-flight preparation and training period. Once all the systems and flight-testing is done and Richard Branson has had a go himself, the first corps of paying-astronauts will climb aboard. When your turn comes, make sure your luggage isn't over the weight limit ... or it could cost you (another) fortune!

TheTerminatorFans.com

TERMINATOR 1984

In the film that launched director James Cameron's galactic career (*Titanic*, *Avatar*) and gave us Arnold Schwarzenegger at his very 'I'll be back' best, a robotic assassin (Schwarzenegger) from a post-apocalyptic future travels back in time to eliminate a waitress (Linda Hamilton), whose son will grow up and lead humanity in a desperate war against sentient machines.

THEME	Military Robots
WHERE	Massachusetts USA
MOVIE TECH	T800 Military Robot
REAL LIFE TECH	Military Robots

Military robots aren't that uncommon, but when they are designed and made by Boston Dynamics, then they are very hard to ignore. The LS3 is a rough-terrain robot designed to go anywhere the Marines and infantry soldiers can go on foot. It helps to carry the soldiers' loads – each robot carrying up to 400lbs of gear – and has enough fuel on board for a twenty-mile mission lasting up to twenty-four hours. Autonomously, the LS3 follows the mission leader using sensors and computer vision, so it doesn't need a dedicated or remote driver. It will also travel to a pre-designated location, using terrain sensing and GPS. LS3 began a two-year field-testing phase in 2012 and is funded by DARPA and the US Marine Corps. Boston Dynamics has assembled an extraordinary team to develop the LS3, including engineers and scientists from Boston Dynamics, Carnegie Mellon, the Jet Propulsion Laboratory, Bell Helicopters, AAI

Corporation and Woodward HRT. More, 'It's got your back' than 'I'll be back', then. It may not be the Terminator Series 800 model we know and love from the film, but we wouldn't like to bump into one of these behemoths in a dark alley!

TERMINATOR 2: JUDGEMENT DAY, TERMINATOR 3: RISE OF THE MACHINES 1991 AND 2003

At the heart of the Terminator sequels sits SkyNet, an artificial intelligence system that has grown self-aware and is determined to protect itself at all costs – even if that includes the extermination of humankind. It operates through fabricated war machines and uses time-travelling cyborgs to neutralize threats to its plans for world domination.

THEME Artificial Intelligence

WHERE New York, USA

MOVIE TECH Skynet

REAL LIFE TECH IBM's 'Watson' Computer

Watson was developed by IBM to demonstrate how clever computers can actually become. It is a computer system that can answer questions put to it in natural language, the first step on the road to developing artificial intelligence. The system was developed at IBM's Deep QA project led by principal investigator David Ferrucci and in 2011, was specifically developed to answer questions on the US quiz show *Jeopardy*! It was pitted against two of the show's previous winners and in an amazing denouement emerged triumphant, winning the one million dollar first prize. During the show, Watson had access to four-terabytes of data, but was not connected to the Internet. In February 2013, IBM announced Watson's first commercial application – making management decisions in lung cancer treatment at a US cancer treatment centre. It has been reported that ninety percent of the staff who use Watson now follow its recommendations and guidance completely. OK, so it's not Skynet, or in reality does it actually think for itself, but it definitely gives credence to the notion that machines are starting to take over!

WHAT DO YOU KNOW ABOUT?
Pollyanna Woodward

Pollyanna is every geek's dream woman: gorgeous and obsessed with tech. She's also an all-action girl who is never afraid to tackle the show's most hair-raising, limb-threatening tech tests. That's why we call it *Polly's tecchie thrills*!

Here are five things you might not know about Pollyanna . . .

1 Right now I am learning to speak German. Vielleicht schaffe ich es, in die deutsche Gadget Show zu kommen.*

2 Believe it or not I am a certified sports masseuse.

3 I absolutely adore theme parks.

4 In between filming The Gadget Show and all the other fly-around-the-word stuff I get asked to do, I have to find time to continue my studies in sports psychology.

5 I love the movies The Goonies and Ferris Bueller's Day Off.

* Translation: 'Maybe I can manage to get on to the German Gadget Show.'

Build Your Own
Jet-Powered Hoverboard

It was the star of *Back To The Future* and every fan of the film would love to have his or her very own hoverboard. But can Jason and a team of experts really manage to build one?

In series five of *The Gadget Show*, Jason created his first version of a hoverboard with a plank of wood and a regular leaf-blower engine. There were technical problems (lack of control over direction mainly), but hover it did. Something everybody on the show agreed was pure 'genius'!

Now it was time to revisit the idea and with the proposal to use an uncomfortably large jet turbine engine to give stability and thrust, it was the moment to call in the experts. Project Thrust was underway . . .

Every successful Gadget Show build begins with a diligent piece of expert design (or in the case of Jason, a quick sketch on the back of an envelope). With the sort of power output envisaged from the jet engine, this was no time to joke around, so in came engineers Simon Oldfield and Stuart Parish.

Jason also called on an old mate, Ali Machinichy, whose expertise in model aircraft powered by gas turbine engines would prove invaluable.

FINISHING TOUCHES ARE PUT TO THE ULTIMATE JET-POWERED HOVERBOARD

Jet engines work on the simple process of suck, bang, blow. Or more accurately, air is sucked in and mixed with fuel in a combustion chamber, the fuel ignites in an explosive reaction and the exhaust gasses are thrown out of the back creating thrust. And the thrust from this particular engine should be more than enough to propel Jason on his board at a not inconsiderable rate of knots.

JASON IS YEARNING TO GET OUT ON TO THE 1.2KM ARMY RUNWAY

So onto the build: First Jason rounded the edges of a plank of marine grade 2.5-metre plywood then drilled holes at either end to take the pipes from the leaf-blowers.

The inflatable underside (think hovercraft) was formed by gluing and stapling durable plastic pond lining to the board and a 1.5m stainless steel grommet fitted in the middle of the board so that when the plastic was inflated it would look more like a doughnut than a half-filled mattress.

THE WIRELESS RADIO HANDSET WILL CONTROL THE THRUST AND PIVOT OF THE JET ENGINE

With the significant upgrade in forward thrust, it was thought vital to install two (rather than the single) leaf-blowers to give enough lift. At over £200 each, the super-powered Stihl BG 86 are light and with four times the power of the blower used on the original build; each one is capable of pumping out 800 cubic metres of air an hour. Almost as much as Jason in full flow.

Of course if these Titans of leaf blowers kept on filling the underside skirt, it would go pop in pretty short order, so to let enough air escape

JASON MAKES STANDING UP ON THE HOVERBOARD LOOK EASY. FOR A FEW SECONDS, ANYWAY

APERFECT END TO A TROUBLESOME DAY

to maintain a solid cushion on which the board (and Jason) could hover, small holes were drilled around the edge of the lining.

Now it was time to fit the engine. Under Ali's supervision, the engine was located centrally at the back, and in an elevated position to provide downward thrust so that Jason would not find himself flying off into the wide blue yonder rather than hovering just above the ground.

To house the 60lb engine, the team would build a steel frame and would need to add two seven-litre fuel tanks. Engineers Simon and Stuart fabricated a pivoting mount to direct the thrust and once all this was bolted on to the board it was almost time to launch The Gadget Show hoverboard Mk2.

With, well, thrust at the heart of Project Thrust, control was going to be vital for the safety of all concerned. Ali programmed a wireless remote control handset so that

Jason would be able to moderate the levels of thrust and angle of the pivoting engine. Finally with the addition of the all-important Gadget Show decals and a quick polish of the shiny metal bits for the photographers, the jet-powered hoverboard was ready to be fired up for the first time.

As Jason tentatively ramped up the power of the jet engine, he couldn't help grinning from ear to ear, shouting above the roar, 'It makes me feel like an evil Bond villain!'

Regular watchers of *The Gadget Show* will know that when a build is out there on the ragged edge of what's feasible, things can and will go wrong. In quick succession rain comes and finds its way into the electrics. The fuel pump shorts out and the hoverboard suddenly looks like a forlorn plank of wood with garden tools bolted to it. With all the engineering skills on hand, it's only a matter of time before things are looking up. The electrics are dried out and rebuilt, the fuel pump replaced and we're ready to try again. Thirty seconds later disaster strikes and the engine pivot control fails. This time considerable ingenuity is called upon and Ali finally resorts to strapping a wooden rudder handle to the engine.

Finally, the hoverboard is up and away. Jason is ecstatic. 'The power of the new hoverboard is way beyond anything I've ever experienced on The Gadget Show.' That says it all really.

Gadget Show Moments That Made Us All Go 'WOW!'

A Fully Immersive Entire Person Shooter.
The Ultimate Battlefield 3 Simulator
Series 16, Episode 11: transmitted 24 October 2011

In one of the show's most amazing builds, a first-person simulator featuring the very latest in entertainment technology was constructed. The sim was housed in an Igloo Vision 360 dome measuring 4m x 9m fitted with five HD projectors that projected *Battlefield 3* onto a colossal wraparound 360-degree screen. Wherever Jason looked he could see the gaming action going on around him! With ten overhead cameras tracking his every move, an omni-directional treadmill meant Jason could move in any direction. The result was fully immersive game play. To increase the player's apprehension, the force of the paintball guns was ramped up making it the only First Person Shooter capable of giving the players genuine battle scars. Finally, former SAS soldier and thriller writer Andy McNabb got in on the action too!

Now That's What I Call a Real RC Car! The Telepresence Car
Series 12, Episode1: transmitted 3 August 2009

Suzi Perry and Jason took on one of the show's most technically challenging tasks ever. They built the best remote control car in the world. Using telepresence technology, the show developed a full-sized road car capable of competing in a proper racetrack-located banger race. Once the presenters were accustomed to the controls, they were able to drive the vehicle around the track from a remote location – keeping themselves out of harm's way in the process. It wasn't easy, but then nothing this cool ever is!

The Lovechild of Jason Bradbury and Darth Vader. Robo Jase Hits the Red Carpet
The World Tour Series 2, Episode 7: transmitted 17 December 2012

From the moment of his birth (as a talking head), Robo Jase has continued to evolve. With the third version wowing crowds at *Gadget show Live Xmas*, Robo Jase is no stranger to the limelight. However, the highest profile event he has ever attended was undoubtedly the Cannes TV festival where he met and spoke to the assembled stars. He could be found on the red carpet itself and on the Promenade de la Croisette while the real Jason hid behind the scenes. Using a sophisticated two-way video and audio link, Robo Jase talked and gestured in sync with the real Jason, becoming the ultimate tech ambassador for T*he Gadget Show*.

To Infinity and Beyond. The Gadget Show Logo in Space
The World Tour Series 2, Episode 1: transmitted 5 November 2012

Jason and Polly went where no Gadget Show presenter has ever gone before – to the edge of space and back again. To be more precise, they sent three tiny high definition action cameras up in a weather balloon to film the stratosphere where temperatures fall to minus 40 degrees centigrade. Once the balloon had deflated, the cameras hurtled back to earth at speeds approaching 300mph. Only the toughest kit could possibly survive the journey. If there was an ultimate test of the durability of action cameras, then surely this was it!

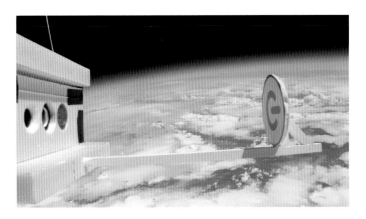

Introducing Jason's Alter Ego, 'Dolphin Man'. The Flyboard 2
The World Tour Series 1, Episode 2: transmitted 30 April 2012

Jason and Polly took to the skies again, only this time with nothing more than the power of pressurised seawater to keep them aloft. The Flyboard 2 took some getting used to, but with a little perseverance and the occasional water-induced tumble, this futuristic and spectacular mode of transportation left both presenters with huge grins on their faces. A unique way to fly, and this is the only H2O powered jetpack on the show that has ever delivered a genuine tech-wow factor.

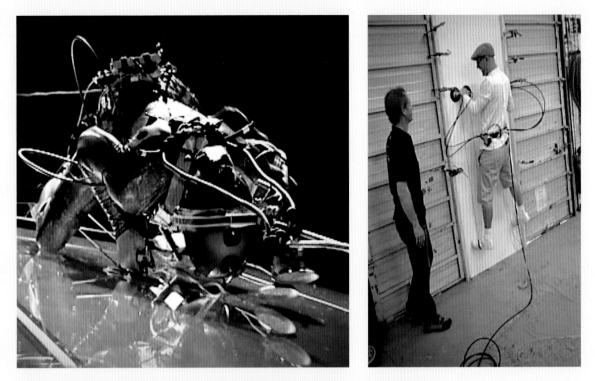

Jason Versus the Big Glass Wall. The Rise of Captain Gecko
Season 12, Episode 8: transmitted 21 September 2009

Jason became a real-life superhero, Captain Gecko. Using nothing more than a green Lycra gecko outfit, nerves of steel, muscle power, purpose built suction cups, a tank of compressed air and a hand-controlled vacuum rig, Jason scaled the glass side of the iconic Fort Dunlop building in Birmingham. Frankly, it was nerve-racking for the production team (including the remote control helicopter pilots) and potentially dangerous for would-be superhero Jason. Captain Gecko emerged triumphant and was ready to save the world another day ... or at least climb up a chunk of it!

On Land or Water? It Really Doesn't Matter. Introducing the Quadski!
Season 18, Episode 1: transmitted 14 October 2013

While visiting the Great Lakes of North America, Polly took advantage of the opportunity to hop on board a piece of hybrid-designed tech that crossed a Jetski with a Quadbike. The Quadski is just as at home on land as it is on the water, and is huge fun to pilot on either. This in-depth test was conducted purely in the name of research you understand, and part of Polly's ongoing search for technological thrills good enough to top her personal leader board. It was a piece of tech the show had been eager to get its hands on for ages, and it didn't disappoint!

Vintage tech ads

'1,000 Songs in your pocket.'

Published circa 2003

Designed chiefly by Susan Alinsangan, Art Director of the Los Angeles and New York based advertising agency TBWA\Chiat\Day, the iPod silhouette campaign ads are among the most iconic tech ads of the last decade. The campaign featured dark silhouetted characters dancing against bright eye catching backgrounds. Each dancer was shown holding an iPod while using Apple's trademark white earphones. In print the ads were visually stunning, while on television Apple added music to make them come alive. Apparently Steve Jobs didn't like the campaign when it was first shown to him, questioning the silhouettes artwork since it failed to present the iPod in much detail or offer much in the way of an explanation of what it did. The ad designers eventually convinced Steve, and copywriter James Vincent added the genius tagline '1,000 songs in your pocket.' that helped clarify things. The bright shades and tropical colours of lime green, yellow, fuchsia, bright blue, and pink were a constant theme, and the white headphones also became an icon signifying the iPod itself. Perhaps the biggest compliment is that the ads even spawned their fair share of parodies in television's *Family Guy* and on YouTube.

★ CLASSICAL ★
Progster

T-34 OUT ON TOUR

VAN HALEN
EXCLUSIVE NEW INTERVIEW BY ALAN TITCHMARSH

TOOL
ART AND PSYCHEDELIA ARE JUST WORDS

ELP
CARL PALMER'S STEEL KIT FINALLY SOLD FOR SCRAP

THE RESURRECTION OF AL MURRAY

JIMMY PAGE
AFTER LED ZEPPELIN IV I JUST WANTED TO BE A GARDENER HE TELLS ALAN TITCHMARSH

Four violinists are Four *too many*

From so nearly being a rock and roll superstar to living in a storage unit, Al Murray tells *Classical Progster* where it all went wrong. And what he's doing about it.

It's an unusually hot day even for the atypical British summers we've been having lately when *Classical Progster* turns up at Al Murray's imposing London residence. We meet in Al's home studio, full of great gear including a rare guitar built by Brian May's dad and a Roland drum kit. There's just about enough room for us to perch on a bunch of packing cases (he is off on an Asian tour at the end of the week) as we explore how life's vagaries meant Al went from almost rock stardom to sleeping in a bass drum, and finally finding salvation through stand-up comedy. Let's start at the beginning.

When and why did you start to play the drums?

I started playing the drums when I was nine years old. When you've seen Animal [the crazed drummer of Dr. Teeth and The Electric Mayhem band from *The Muppet Show*] on the TV and caught a glimpse of the life he led and the lifestyle he had, it seemed to be the only option. The drugs, the women, the fame and the big drum kits. Especially the really big drum kits.

And at nine did you get a big drum kit?

No, I started off at nine in the classic manner with pots and pans and then moved on to my first snare drum. I was given some snare drum exercises to do, didn't do them and then found a drum kit at school.

What was school like for you?

Well, when you are burning with the fire of hot rock and roll, it is very difficult to fit in.

Did you find any kindred spirits at school?

Yeah! I ended up in a jazz-fusion band with a guy called Dicky Weatherall playing keyboards and a guy named Ned Boulting, who some people may know from the Tour de France on telly, playing saxophone.

Why jazz-fusion?

It was sort of Chick Korea kinda stuff and was the only option at the time because we weren't allowed to form a rock band at school. They didn't think it was proper music, so a jazz band was the only outlet. We did end up performing our own lukewarm fusion. There was a big band, a swing band, but no opportunity to put a rock band together or rehearse one. Not even in secret. I mean nothing was secret at boarding school.

What happened next?

Then I went to university and fell in with a band with a guitar player called Simon Oakes and we tried to square the circle of shoe-gazing indie, progressive rock and heavy metal. We wanted to combine all three elements in a band.

And did it work?

It depends on whom you ask.

Well it must have had some merit because you got picked up by a major label pretty quickly.

Well, we were called The Captain Feedback Invitation Orchestra at the time because we had a revolving line up – several bass players in and out; a real

Al and the members of Geyser laying down the rhythm track for *It Never Snows in September* at Abbey Road studios. It would be the break through B-side that launched the band's career.

revolving door of bass players. We tried a double drummer line up but that didn't work because they couldn't fit around my playing. We had a female singer – who I am ashamed to say was there purely for decoration. Then Baggy[1] happened and we somehow missed the boat on that.

How did you progress to becoming the nearly world famous professional musician?

After that The Captain Feedback Invitation Orchestra slimmed itself down to Captain Feedback. We played a lot of gigs with a lot of interest on the Camden scene. At one point I think we opened for a proto-Radiohead and at one time for proto-Blur. But I am not sure, as their haircuts have changed since.

Where you managed at this point?

We managed, yes.

---

1 Baggy, so named after the loose-fitting clothes favoured by bands and fans alike, was a dance-oriented style of rock music that was much influenced by the Madchester scene in the 1980s. It featured psychedelic-type guitar sounds coupled with a funky drumbeat.

Al tunes up the Red Special. Later he would use a super-sized Roland drum set that would cause problems on tour and help sow the seeds of discontent within the band.

No, when did you get professionally managed?

By now the band had mutated again, at this point, into a band called Geyser – but that's a geyser like a hot spring, not like a bloke, which is what I have to say every time the name of the band comes up. It's one of those band names that doesn't quite make it because you have to explain it every time you say it. I expect

Have you discussed this specifically with Bono?

No, I haven't but I will the next time I see him. The last time I say him was at the Q Awards the year before last, and he asked me if I was going to have my picture taken with band, I should bend my knees so I wouldn't look too tall. Or the band too short, which they are. Haven't spoken to him since.

'We tried to square the circle of shoe-gazing indie, progressive rock and heavy metal . . .'

U2 for a long time before they became famous kept saying, "No it's a 'U' and '2' like the airplane."

So we had changed into Geyser at this point and we managed ourselves; so there was the creative tension, the

'I personally think music should have stopped around about 1981.'

administrative tension and then the simple personal tension of being probably the most exciting three-piece rock band on the London circuit in 1994 and early 1995.

What happened in late 1995 because suddenly there is a huge buzz and you get scaled up to stadium rock? How did it all come about?

Well, it was luck really. One of our B-sides, Tool[2] took a liking to it. I think it was Tool anyway, I'm not sure, maybe it wasn't Tool, in fact I don't think it was Tool, but that's not the point. Anyway they used to play it before they came on in Japan. And the record got picked up in Japan, turned into an A-side and the next thing we knew we were doing support slots all over the world.

And this B-side was recorded in Abbey Road?

Yeah!

And you wrote it?

That's still something we haven't actually usefully settled legally. I can't really talk about who wrote it. It depends which pressing you get. On each different pressing of the record there's a different credit on it. It might as well have been written by my lawyer.

What was the track called?

That tune was called *It Never Snows in September*.

So the success of the record means you are out on the road, and, of course, your first port of call is the Australian tour – where it went very well at first.

You only ever have a good time in Australia, that's the simple truth; they're nice people, the weather is good, the recreational opportunities are excellent – it's the perfect place to go if you are a rock band on the way up. It was the second trip that was the problem.

So what happened on the second trip?

OK, we had decided to actually hire some violinists. On one of the second album tracks there is a middle eight. I say middle eight it was more like a middle sixty-four because it was one of those, you know, *magnum opus* things. *Why Wait?* was the name of the tune which was the whole of the B-side of the album. I am still talking about B-sides. And we got some violinists in for that and basically, it turned out that they were on some sort of Australian musicians' union rates – so we've spent our whole tour budget on four violinists. It wasn't my idea; I would have used a synth.

And this is what caused all the problems?

This is where it fell apart. Suddenly the money vanishes, we're on the other side of the world, we're there in winter when

2 Tool are an American Grammy-award winning progressive art-rock band formed in Los Angeles in 1990 and still touring today.

the weather's not so great and all the good things about touring Australia had evaporated.

What did you do?

I came home but the bass player stayed and that was the main fragment in the band. He went to live with the four violinists. Up the Murray River, somewhere.

So you come back and try to reorganize the band, get a new bass player, create a new sound?

Yeah, I love bass players and it's important to have a good relationship with one, but you've got to find one to have a good rela-

I was fired from the band. That's when it went wrong for me basically because him and the guitar player then teamed up.

But it didn't go entirely wrong for you. You managed to jump ship and spend time playing with some pretty impressive names.

I can't really say who they all are though because, again, when you work with some of the names we could talk about, in the kind of Smashing Pumpkins kind of area . . . well, I went to America for a bit, had to come back. I hadn't sorted my work permit out. What they were doing at the time, that band, was bringing Brit-

'That's still something we haven't actually usefully settled legally. I can't really talk about who wrote it. It depends which pressing you get.'

tionship with. So we auditioned people for, I think, six months and all the time we're running out of money, there's an ongoing royalties dispute and in the end we replaced the bass player with a Roland sequencer, which seemed to be the sensible option.

So you are down to a two-piece now?

We were a two-piece as you say, but when the Beatles got in Billy Preston, George Harrison did it because he wanted someone around so they would behave themselves, and I decided to bring in a friend of mine called Mike to play keyboards. And that was the point when

ish guys over, and if they liked you, they'd sort out your work permit and if they didn't, they would have you deported. It's a tough world, rock and roll.

Especially if the police get involved!

Yeah, exactly! Well I did play with them for a bit but that wasn't much fun . . . Stewart Copeland, there's only room for one drummer in the Police.

And sadly for you that was Stewart?

Yeah, Stewart, of course!

So you're now back in the UK and I gather by now your living circumstances were pretty rough. The drum kit had been pawned?

Well, one of the drum kits has been pawned. The really big one, the one I had in Australia. That was one of the other issues in Australia – it wasn't just the four violinists. It was the fact that I required my own two artics to shift the drums.

Back on the road again after all these years. T-34 live on stage at the Skopje Jazz Festival in Macedonia, where they played to a sometimes-bemused audience.

That sounds very Carl Palmerish. There were rumours that the infamous ELP tour with the steel kit that needed a crane to lift it on stage was something you were keen to emulate?

The thing is you don't really need six tympani, but if you are going to get them given to you by a drum company, then you're going to ship them to Australia and they're going to have their own lorry and they're going to have their own sound engineer and monitor mixer. They are going to have all those things and why not?

And why not indeed?

So anyway, I pawned that kit but my other kit, the original, the Red Special as

I called it, that was in storage. So I ended up living in storage with the drum kit. Luckily there was a pillow in the bass drum, so it worked out OK.

Were you determined to get back into music?

I had a year off from it, a year away from it. Didn't even listen to music.

If not music, then what were you doing?

That was when I first wondered about doing stand up comedy, but you know my mind wasn't made up on that. It looked like a better way of making money because there is not four of you, or three of you or two of you. Or three of you and four violinists.

So you tried the stand up comedy route. How did it come about? I mean the rumour at the time was that you were permanently drunk in pubs and people couldn't shut you up.

Well, sometimes rumour isn't far off the truth. I don't want to glorify the fact that I turned that desperate situation into something that has worked out very nicely for me. That would be to encourage others to hit rock bottom. But yeah, maybe.

'When Led Zeppelin IV exists there is no point in anyone else picking up a guitar.'

But you are back on the music scene now?

Yeah! The T-34 is my current project. The thing is, what I realised about rock music is that there is really no point in writing any more music. I personally think music should have stopped around about 1981. There's enough of it they could just endlessly reissue; no need to pay any new musicians, no need to have any new studios or anything. Stop music then. And

I think all the great rock music has been written – so the T-34 that I play with now, well we play the standards, old classics, the original tunes, the source of rock and roll as it should be. Led Zeppelin, the Kinks, a bit of The Who, the Stooges. The real deal! I look at the stuff we were doing in the nineties and I regret every single note we ever played. It was a red herring. When *Led Zeppelin IV* exists there is no point in anyone else picking up a guitar.

Yeah, I think Jimmy Page said that once didn't he?

He definitely did say that, which asks a lot of questions why he made any albums afterwards. But there you go.

Got to earn a living I suppose?

I suppose, yeah.

And so now it's T-34 and you are touring again?

Yes, we do an annual world tour of one show because the thing is when you are that dedicated to going to the actual source of rock and roll music, you can't spread it too thin. If you play the O2 Arena then you are spreading it too thin. if you play five O2 Arenas then you are spreading it way too thin. What you want to do is boil it down; it's like a sauce – you want cook it down to the purest reduction of rock and roll. So we play one gig a year. Very exclusive and tickets are a hundred quid each. What we're really interested in is keeping the flame alive. If you look at what's going on out there in pop music – auto-tune, drum machines – it's horrible. If we want to sing out of tune, we will.

It's been a hard road back, but Al is happiest rediscovering his paradiddle.

Intentionally?

For sure! And If I end up playing out of time, that's because I am a human being. I'm not some robot. You know this isn't the drone warfare of rock and roll; it's the real thing!

And T-34 are performing at the back end of this year, I gather?

Yes, we'll hopefully fit a show in before the end of the year.

Well thanks Al. I hope all continues to go well for you.

Al Murray is patron of a charity called Cambodian Children's Charity (www.camkids.org).

It is involved in urban and rural projects getting kids into school, providing medical and dental care, building schools and all sorts of stuff. A golden pound organization, all the funds raised go straight to those who need them.

AND THE REST
ULTIMATE GADGET FACE-OFFS

So what about all the other pieces of iconic tech that have taken their place in gadget history? Whether used for domestic purposes like the Hoover or for fun like Hornby model railways, most have become commonplace, even part of our everyday lives.

We all have our personal favourites (Jason's is the Sinclair C5!) and each has its own set of interesting facts and attributes. For many reasons this will be the hardest set of tech v. tech comparisons to make. Rest assured there will be winners, but the battles will be close fought affairs!

SCALEXTRIC

Success: 4
Scalextric was so popular on launch that the parent company was unable to meet demand and was sold to Line Bros Ltd, who operated as Tri-ang.

Lifetime: 4
1958 to now.

Innovations: 4
Changed model cars from simple wind-up toys that you had to retrieve from under the sofa to cars that would stick to a track and whose throttle was controlled electronically. Races were now practical and fun.

Fun Facts
In 2001, five people were recorded as being hospitalised through Scalextric-related injuries.

FACEBOOK

Success: 5
1.26 billion active users (June 2013). Total number of photos uploaded now stands at a staggering 250 billion and is growing by 350 million a day!

Lifetime: 4
2006 to now.

Innovations: 5
Facebook is now a fully integrated component in the lives of billions, and has taken social networking to a new level of sophistication with special apps and integration with other websites.

Fun Facts
Before Facebook, its creator Mark Zuckerberg launched 'Facemash' – a site that compared two people's faces and asked which was hotter!

HORNBY

Success: 5
Since its first appearance, Hornby has been the leading brand of model railways in the UK.

Lifetime: 5
1920 to now.

Innovations: 4
One of the first ever manufacturers of model railways who have kept on innovating, notably when they introduced electric models to replace clockwork trains.

Fun Facts
The first commercially produced 00 guage live steam locomotive was launched by Hornby in September 2003.

MYSPACE

Success: 4
From 2005 until early 2008, Myspace was the most visited social networking site in the world.

Lifetime: 3
2003 to now, but largely overtaken by Facebook.

Innovations: 4
The first mainstream popular social networking website.

Fun Facts
Myspace is known for having introduced Lilly Allen to the world.

ROUGH RIDER

Success: 5

Hundreds of thousands of Rough Riders were sold in the first few years of production. It was recently re-launched as the 'Buggy Champ' 30th anniversary edition.

Lifetime: 4

1979 to now.

Innovations: 4

The world's first truly mass-produced off-road RC car.

Fun Facts

Manufacturer Tamiya gave Rough Riders to leading Grand Prix drivers such as James Hunt, who would then be seen playing with them at race weekends.

GOCYCLE

Success: 4

Originally released in 2009, it quickly sold out. By 2011 the model had been upgraded. Gocycle was nominated for Commuter Gadget of the Year at the 2009 T3 Gadget Awards.

Lifetime: 3

2009 to now.

Innovations: 4

One of the first popular, successful, and even iconic electric bikes. It was the brainchild of former McLaren F1 industrial designer, Richard Thorpe.

Fun Facts

Gocycle is the world's first production bicycle to feature a completely enclosed multi-speed chain drive with side-mounted wheels, which means no grease or grime on your clothes and makes fixing a flat tyre easy.

DYSON

Success: 4

Within eighteen months of its launch, the Dyson became the best-selling cleaner in the UK.

Lifetime: 4

1993 to now.

Innovations: 3

The first bagless vacuum cleaner was quickly followed by the first with a ball design. Perhaps the first vacuum cleaner to attain iconic design status.

Fun Facts

Sir James Dyson went through 5,127 prototypes, before he invented the first cyclonic bagless vacuum cleaner.

VENESSA'S LUNCHBOX

Success: 4

One of the best-selling RC cars of all time and was even re-launched in 2005 because of its enduring popularity.

Lifetime: 5

Originally released in 1987, in 2008 it was still the biggest selling RC car from Tamiya.

Innovations: 3

The model was based on monster trucks with huge tyres of the 1980s, something no other manufacturer had thought to do. It could also wheelie successfully!

Fun Facts

Original packaging artwork included the phone number of the sister of a Tamiya employee who received loads of calls from people thinking she was Venessa.

SINCLAIR C5

Success: 1

Disastrous. Only somewhere in the region of 12,000-17,000 units were ever sold.

Lifetime: 0

It launched on 10 January 1985, but production ceased later that year in August. The company went into receivership on 12 October 1985. It took just eight months to go from release to oblivion!

Innovations: 4

One of the first electric commuter vehicles.

Fun Facts

The C5's top speed of 15mph (24km/h) was chosen deliberately as this meant a driver wouldn't need a driver's license. Even disqualified drivers could take one out on British roads.

HOOVER

Success: 4

Originally invented in 1908, and by the 1960s it was claimed that there was a vacuum in virtually every home in the US.

Lifetime: 5

1908 to now.

Innovations: 4

The first ever effective and useable vacuum cleaner.

Fun Facts

The Hoover was actually invented by James Murray Spangler who showed a working model to his cousin Susan Hoover. Susan's husband and their son acquired the patent and the rest is domestic appliance history.

Vintage tech ads

'The heart of a system that grows with you.'

Published circa 1981

Available as a kit for £49.95 or fully built for £69.95, the Sinclair ZX81 made the personal computer accessible to all and its inventor, Sir Clive Sinclair, created a whole new market by enticing people who would have never previously considered owning such an arcane device. At the time the Apple II Plus Personal Computer cost £630, so the ZX81 was arguably the computing bargain of the decade. Initially, it could only be bought direct from Sinclair; however, the ZX81 would go on to sell a staggering one and a half million units in just three years, helped latterly by a 112-store distribution deal with WH Smith. Sold alongside computer software and blank cassette tapes, the ZX81 was the centrepiece of the stores' 'Computer Know-How' section.

Sinclair ZX81 Pe
the heart of a sy

you
plug
the
data

You'll find the ZX81 an ideal introduction to the world of computing. And as your skills and needs develop, so your ZX81 system keeps pace.

or a
little
add

You can add 16-times more memory with the ZX 16K RAM.

The exciting ZX Printer is now available.

And the range of Sinclair ZX Software is growing all the time.

som
– th
syst

Our latest product is the ZX81 Learning Lab described opposite.

To order any of these products, simply use the no-stamp-needed order form on the back of this leaflet.

al Computer-
m that grows with you.

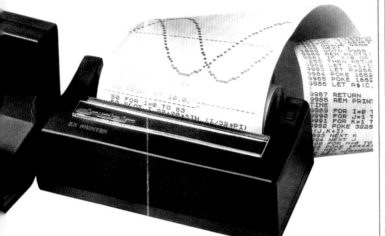

ZX81 Learning Lab.

The ultimate course in ZX81 BASIC programming.

Some people prefer to learn their programming from books. For them, the ZX81 BASIC manual is ideal.

But many have expressed a preference to learn *on* the machine, *through* the machine. Hence the new cassette-based ZX81 Learning Lab.

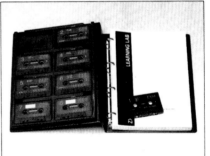

oyte RAM pack for ive add-on memory.

a complete module to fit 31, the RAM pack simply isting expansion port at omputer to multiply your orage by 16!

g and complex programs database. Yet it costs as orice of competitive ory.

M pack, you can also run e sophisticated ZX Software Household management mple.

The ZX Printer- for only £59.95.

Designed exclusively for use with the ZX81, the printer offers full alphanumerics and highly sophisticated graphics.

A special feature is COPY, which prints out exactly what is on the whole TV screen without the need for further instructions.

At last you can have a hard copy of your program listings – particularly useful when writing or editing programs.

And of course you can print out your results for permanent records or sending to a friend.

Printing speed is 50 characters per second, with 32 characters per line and 9 lines per vertical inch.

The ZX Printer connects to the rear of your computer – using a stackable connector so you *can* plug in a RAM pack as well. A roll of paper (65 ft long x 4 in wide) is supplied, along with full instructions and an order form for more paper.

The package comprises a 160-page manual and 8 cassettes. 20 programs, each demonstrating a particular aspect of ZX81 programming, are spread over 6 of the cassettes. The other two are blank practice cassettes.

The course requires absolutely no previous computer experience – even the words are explained as you come to them. And although there are sections on the 16K RAM pack and ZX Printer, all you need is a standard Sinclair ZX81.

GADGET SHOW LIVE 2013

2013 saw the fifth anniversary *Gadget Show Live* at Birmingham's NEC – the most eagerly anticipated tech event of the year. The crowds were enormous, the gadgets and tech on display mind-boggling and, of course, right at the heart of it as always was the hour-long Super Theatre live show itself.

It's a huge undertaking to put on a theatrical performance with full audience participation in front of a five thousand-strong crowd, and the behind-the-scenes technicians, production team and presenters spend a very long day rehearsing to get everything just right.

Here's our look at just what goes into making *Gadget Show Live* the best tech show in town!

BACKSTAGE

Construction of the *Gadget Show Live* stage is almost complete.

The behind-the-scenes technical team making sure everything is working.

Naomi, our resident dancer, gets ready for the show.

Jon Bentley puts on his microphone and waits for his stage call.

Stage crew in the wings make sure that all the presenters' cues are correctly sequenced.

Backstage crew taking a much needed break during the long hours of rehearsal.

Five microphones (and an apple?) are laid out ready for the presenters.

Gareth, as soundman at the back of the theatre, has an important job to do.

Polly being her usual giggly self.

Doctor Dom Sagolla explains how the mind-controlled skateboard functions.

Gadget Show Live dancers looking wonderfully graceful during rehearsals before the show.

Gadget Show producer Ewan Keil is in charge of rehearsals. What is it about Polly and watermelons?

The presenters get ready for some mind-controlled watermelon exploding. Polly is determined to wind up the crew!

Jason demonstrates how the audience will be able to text the presenters during the show.

On the big screen Dom is about to blow up a watermelon.

Jason explains the Airport Race Game to the crowd.

Jason chats with two Robo Challenge engineers, rehearsing
how he will interact with audience members during the show.

Time to try out some planned audience participat[...]
Jason is not overly keen on this segment of the s[...]

Jon looks deceptively relaxed on stage although he will spend
all afternoon desperately trying to learn his lines!

Polly always loves an audience!

The Trapeze Artists' performance was
designed to demonstrate the picture
quality of the GoPro action camera.

Polly is ready for the personal
transport race on her Ecospin.

No *Gadget Show Live* would be
complete without Titan The Robot.

Jason tries out the EDF Jetpack.

The stage is built, sound and lights tested, the script written and edited, and the presenters have rehearsed for a day. It's time to lift the curtain on *Gadget Show Live* 2013.

A FEW MOMENTS FROM THE SHOW THE NEXT DAY

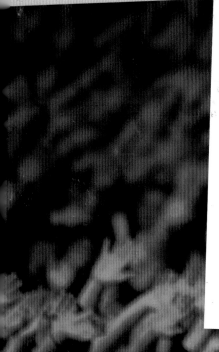

Jason and his team (one third of the audience) play the Airport Race Game.